T0142826

Studies in Computational Intelligence

Volume 727

Series editor

Janusz Kacprzyk, Polish Academy of Sciences, Warsaw, Poland
e-mail: kacprzyk@ibspan.waw.pl

About this Series

The series "Studies in Computational Intelligence" (SCI) publishes new developments and advances in the various areas of computational intelligence—quickly and with a high quality. The intent is to cover the theory, applications, and design methods of computational intelligence, as embedded in the fields of engineering, computer science, physics and life sciences, as well as the methodologies behind them. The series contains monographs, lecture notes and edited volumes in computational intelligence spanning the areas of neural networks, connectionist systems, genetic algorithms, evolutionary computation, artificial intelligence, cellular automata, self-organizing systems, soft computing, fuzzy systems, and hybrid intelligent systems. Of particular value to both the contributors and the readership are the short publication timeframe and the worldwide distribution, which enable both wide and rapid dissemination of research output.

More information about this series at http://www.springer.com/series/7092

Roger Lee

Editor

Applied Computing & Information Technology

Editor
Roger Lee
Software Engineering and Information
   Institute
Central Michigan University
Mt. Pleasant, MI
USA

ISSN 1860-949X   ISSN 1860-9503   (electronic)
Studies in Computational Intelligence
ISBN 978-3-319-87709-9   ISBN 978-3-319-64051-8   (eBook)
DOI 10.1007/978-3-319-64051-8

© Springer International Publishing AG 2018
Softcover reprint of the hardcover 1st edition 2017
This work is subject to copyright. All rights are reserved by the Publisher, whether the whole or part
of the material is concerned, specifically the rights of translation, reprinting, reuse of illustrations,
recitation, broadcasting, reproduction on microfilms or in any other physical way, and transmission
or information storage and retrieval, electronic adaptation, computer software, or by similar or dissimilar
methodology now known or hereafter developed.
The use of general descriptive names, registered names, trademarks, service marks, etc. in this
publication does not imply, even in the absence of a specific statement, that such names are exempt from
the relevant protective laws and regulations and therefore free for general use.
The publisher, the authors and the editors are safe to assume that the advice and information in this
book are believed to be true and accurate at the date of publication. Neither the publisher nor the
authors or the editors give a warranty, express or implied, with respect to the material contained herein or
for any errors or omissions that may have been made. The publisher remains neutral with regard to
jurisdictional claims in published maps and institutional affiliations.

Printed on acid-free paper

This Springer imprint is published by Springer Nature
The registered company is Springer International Publishing AG
The registered company address is: Gewerbestrasse 11, 6330 Cham, Switzerland

Foreword

The purpose of the 5th International Conference on Applied Computing and Information Technology (ACIT 2017) held on July 9–13, 2017 in Hamamatsu, Japan is to bring together researchers, scientists, engineers, industry practitioners, and students to discuss, encourage and exchange new ideas, research results, and experiences on all aspects of applied computers and information technology, and to discuss the practical challenges encountered along the way and the solutions adopted to solve them. The conference organizers have selected the best 12 papers from those papers accepted for presentation at the conference in order to publish them in this volume. The papers were chosen based on review scores submitted by members of the program committee and underwent further rigorous rounds of review.

In Chapter "Meal-assistance Robot Operated by Head Movement", Hiromasa Tomimoto, Shingo Aramaki, Shota Nakashima, Shenling Mu, Kazuo Haruyama, and Kanya Tanaka propose a meal assistance robot with maneuverable interface with head movement is proposed for upper-limb-crippled people. The proposed robot with orthogonal structure, that offers simple and stable motions, is able to provide foods to users smoothly. Thus, the system in compact size can complete the food providing effectively.

In Chapter "Decision Tree Analysis in Game Informatics", Masato Konishi, Seiya Okubo, Tetsuro Nishino, and Mitsuo Wakatsuki extract features of the player program through decision tree analysis. The features of programs are extracted by generating decision trees based on three types of viewpoints.

In Chapter "Binary Blockchain: Solving the Mining Congestion Problem by Dynamically Adjusting the Mining Capacity", Yoohwan Kim and Juyeon Jo propose a novel parallel mining method that can adjust the mining capacity based on the transaction load. They describe how and when to divide or merge blockchains, how to solve the imbalanced mining problem, and how to handle the difficulty levels and rewards.

In Chapter "An Efficient Signature Scheme for Anonymous Credentials", Chien-Nan Wu, Chun-I Fan, Jheng-Jia Huang, Yi-Fan Tseng, and Hiroaki Kikuchi use a partially blind signatures and chameleon hash functions to propose a signature scheme such that both the prover and the verifier achieve efficient authentication to overcome the drawback of anonymous credential systems.

In Chapter "Improve Example-Based Machine Translation Quality for Low-Resource Language Using Ontology", Khan Md Anwarus Salam, Setsuo Yamada and Nishio Tetsuro propose to use ontology to improve the performance of an EBMT system for low-resource language pair. The EBMT architecture use chunk-string templates (CSTs) and unknown word translation mechanism.

In Chapter "A Fast Area Labeling Method Using Auxiliary Lines", Noboru Abe, Kohei Kuroda, Yosuke Kamata, and Shogo Midoritani propose a real-time method to determine label positions using the intersections of the auxiliary and boundary lines of a given area. Experimental results demonstrate that the proposed method is capable of real-time processing and can determine effective label positions.

In Chapter "Heuristic Test Case Generation Technique Using Extended Place/Transition Nets", Tomohiko Takagi, Akinori Akagi, and Tetsuro Katayama show a novel heuristic test case generation technique using an extended PN (place/transition net). A guard and action are introduced to improve the representation power of a PN. Also, a weight that represents the degree of testing priority is given to each transition of a PN.

In Chapter "Risk Assessment of Security Requirements of Banking Information Systems Based on Attack Patterns", Krissada Rongrat and Twittie Senivongse present an initial risk assessment method to assist the project team in validating security requirements of a banking information system. They evaluate the performance of security compliance checking in terms of F-measure and accuracy, and validity of risk assessment in terms of correlation with security expert judgment.

In Chapter "mCITYPASS: Privacy-preserving Secure Access to Federated Touristic Services with Mobile Devices", Macià Mut-Puigserver, M. Magdalena Payeras-Capellà, Jordi Castellà-Roca, and Llorenc Huguet-Rotger present an electronic ticketing system intended to be used for touristic services. Their proposal is the first one that can be implemented on portable devices, such as smartphones, and is flexible enough to include reusable and non-reusable services in the same citypass.

In Chapter "Heuristic-Based Usability Evaluation Tool for Android Applications", Kwandee Phetcharakarn and Twittie Senivongse attempt to assist the evaluators by automating the evaluation task against a number of design heuristics. The paper presents a development of a usability evaluation tool for Android applications by inspecting source code and reporting locations in the code where violations of heuristics are found.

In Chapter "Automated Essay Scoring System Based on Rubric", Megumi Yamamoto, Nobuo Umemura, and Hiroyuki Kawano propose an architecture of automated essay scoring system based on rubric, which combines automated scoring with human scoring. Rubrics are valid criteria for grading students' essays. Their proposed rubric has five evaluation viewpoints "Contents, Structure, Evidence, Style, and Skill" and 25 evaluation items which are subdivided viewpoints.

In Chapter "Mobile Development Tools and Method Integration" Mechelle Grace Zaragoza, Roger Y. Lee, and Haeng-Kon Kim show how a similar set of principles, practices, and programming tools can be combined with recent work by MetaObjects to provide a framework for methods and tools as well as configuration.

It is our sincere hope that this volume provides stimulation and inspiration, and that it will be used as a foundation for works to come.

July 2017

Takaaki Goto
Ryutsu Keizai University, Japan

Contents

Contributors

Noboru Abe Faculty of Information and Communication Engineering, Osaka Electro-Communication University, Neyagawa-Shi, Osaka, Japan

Akinori Akagi Graduate School of Engineering, Kagawa University, Takamatsu-Shi, Kagawa, Japan

Shingo Aramaki Graduate School of Sciences and Technology for Innovation, Yamaguchi University, Yamaguchi, Japan

Jordi Castellà-Roca Department d'Enginyeria Informàtica i Matemàtiques, UNESCO Chair in Data Privacy, Tarragona, Spain

Chun-I Fan Department of Computer Science and Engineering, National Sun Yat-sen University, Kaohsiung, Taiwan

Kazuo Haruyama Department of Electrical Engineering, National Institute of Technology, Ube College, Ube, Japan

Jheng-Jia Huang Department of Computer Science and Engineering, National Sun Yat-sen University, Kaohsiung, Taiwan

Llorenç Huguet-Rotger Department de C. Matemàtiques i Informàtica, Universitat de les Illes Balears, Palma, Spain

Juyeon Jo Department of Computer Science, University of Nevada Las Vegas, Las Vegas, NV, USA

Yosuke Kamata Faculty of Information and Communication Engineering, Osaka Electro-Communication University, Neyagawa-Shi, Osaka, Japan

Tetsuro Katayama Institute of Education and Research for Engineering, University of Miyazaki, Miyazaki-Shi, Miyazaki, Japan

Hiroyuki Kawano Faculty of Science and Engineering, Nanzan University, Nagoya, Japan

Hiroaki Kikuchi Department of Frontier Media Science, Meiji University, Surugadai, Japan

Haeng-Kon Kim Catholic University of Daegu, Gyeongsangbuk-do, South Korea; Department of Computer Science, University of Nevada Las Vegas, Las Vegas, NV, USA

Yoohwan Kim Department of Computer Science, University of Nevada Las Vegas, Las Vegas, NV, USA

Masato Konishi Graduate School of Informatics and Engineering, The University of Electro-Communications, Chofu, Japan

Kohei Kuroda Faculty of Information and Communication Engineering, Osaka Electro-Communication University, Neyagawa-Shi, Osaka, Japan

Roger Y. Lee Computer Science Department, Central Michigan University, Mount Pleasant, USA

M. Magdalena Payeras-Capellà Department de C. Matemàtiques i Informàtica, Universitat de les Illes Balears, Palma, Spain

Shogo Midoritani Faculty of Information and Communication Engineering, Osaka Electro-Communication University, Neyagawa-Shi, Osaka, Japan

Shenling Mu Graduate School of Science and Engineering, Ehime University, Matsuyama, Japan

Macià Mut-Puigserver Department de C. Matemàtiques i Informàtica, Universitat de les Illes Balears, Palma, Spain

Shota Nakashima Graduate School of Sciences and Technology for Innovation, Yamaguchi University, Yamaguchi, Japan

Tetsuro Nishino Graduate School of Informatics and Engineering, The University of Electro-Communications, Chofu, Japan

Seiya Okubo School of Management and Information, University of Shizuoka, Shizuoka, Japan

Kwandee Phetcharakarn Department of Computer Engineering, Faculty of Engineering, Chulalongkorn University, Bangkok, Thailand

Krissada Rongrat Department of Computer Engineering, Faculty of Engineering, Chulalongkorn University, Bangkok, Thailand

Khan Md Anwarus Salam IBM Research, Tokyo, Japan

Twittie Senivongse Department of Computer Engineering, Faculty of Engineering, Chulalongkorn University, Bangkok, Thailand

Tomohiko Takagi Faculty of Engineering, Kagawa University, Takamatsu-Shi, Kagawa, Japan

Kanya Tanaka Graduate School of Science and Engineering, Meiji University, Tokyo, Japan

Nishio Tetsuro Graduate School of Informatics and Engineering, The University of Electro-Communications, Tokyo, Japan

Hiromasa Tomimoto Department of Electrical and Computer Engineering, National Institute of Technology, Gifu College, Motosu, Japan

Yi-Fan Tseng Department of Computer Science and Engineering, National Sun Yat-sen University, Kaohsiung, Taiwan

Nobuo Umemura School of Media and Design, Nagoya University of Arts and Sciences, Nagoya, Japan

Mitsuo Wakatsuki Graduate School of Informatics and Engineering, The University of Electro-Communications, Chofu, Japan

Chien-Nan Wu Department of Computer Science and Engineering, National Sun Yat-sen University, Kaohsiung, Taiwan

Setsuo Yamada NTT Corporation, Tokyo, Japan

Megumi Yamamoto School of Contemporary International Studies, Nagoya University of Foreign Studies, Nisshin, Japan

Mechelle Grace Zaragoza Catholic University of Daegu, Gyeongsangbuk-do, South Korea

Meal-Assistance Robot Operated by Head Movement

Hiromasa Tomimoto, Shingo Aramaki, Shota Nakashima,
Shenling Mu, Kazuo Haruyama and Kanya Tanaka

Abstract The number of crippled people is growing in Japan. They are facing many kinds of difficulties in their daily lives. Especially, it is even difficult in having meals for upper-limb-crippled people. They need system which is able to help them in having meals. There are already several types of robots for meal assist with interface operated with hand originally. The upper-limb-crippled people needs interface with simple operation. Therefore, in this paper a meal assistance robot with maneuverable interface with head movement is proposed for upper-limb-crippled people. The head movement is designed to be detected by a Kinect sensor connected to control unit based on Raspberry Pi. In addition, the proposed robot with orthogonal structure, that offers simple and stable motions, is able to provide foods to users smoothly. Thus, the system in compact size can complete the food providing effectively.

Keywords Meal assistance robot · Upper-limb-crippled people · Operation interface · Head movement

H. Tomimoto (✉)
Department of Electrical and Computer Engineering, National Institute of Technology,
Gifu College, Motosu, Japan
e-mail: tomimoto@gifu-nct.ac.jp

S. Aramaki · S. Nakashima
Graduate School of Sciences and Technology for Innovation, Yamaguchi University,
Yamaguchi, Japan

S. Mu
Graduate School of Science and Engineering, Ehime University, Matsuyama, Japan

K. Haruyama
Department of Electrical Engineering, National Institute of Technology, Ube College,
Ube, Japan

K. Tanaka
Graduate School of Science and Engineering, Meiji University, Tokyo, Japan

© Springer International Publishing AG 2018
R. Lee (ed.), *Applied Computing & Information Technology*,
Studies in Computational Intelligence 727, DOI 10.1007/978-3-319-64051-8_1

1

1   Introduction

In Japan, the number of crippled people is growing in recent years [1, 2]. The growth is considered to continue in near future owing to the background of aging society. The crippled people are facing severe situation with many kinds of difficulties in their daily lives. Especially, for the upper-limb-crippled people, it is even difficult to have meals by themselves. They have to get assistance from their family or helpers in eating. It is a frustrated situation for the care receivers who cannot complete basic daily activities. Meanwhile, it is also a heavy work for their families or care centers. To help the care receivers in eating, it is important to match the pace and amount of help giver and receiver. It is such a careful work that the meal assistance requires at least one helper in a meal in home nursing or at care centers. Considering the aging society in Japan with shortage of labors, it is not an easy problem to solve. According to the situation stated above, some robots are introduced to help the upper-limb-crippled people in eating, relieving the severe situation.

Handy1 [3–5], iARM [6] and My-spoon [7, 8] are well known as robots which are able to assist in eating. Handy1 is developed by Rehab-Robotics Company Ltd in United Kingdom. It is a robot system to provide support in daily life of upper-limb-crippled people. This system helps in eating by spoon up food to users. By that way, the robot is suitable for helping to have liquid food particularly. However, it has difficulty in assistance to have solid-form food which is not easy to spoon up. iARM is developed by Exact Dynamics in Netherlands. It is a robot arm that is designed to help in all living activities. The robot arm is operated by using a joystick. For this reason, it is difficult to operate the iARM for upper-limb-crippled people. Meanwhile, in Japan, there is a dedicated robot system developed to assist eating. It is named as My-spoon developed by Secom Co., Ltd. The system specializes in assistance in eating than the two robots introduced above. The meal assistance robot employs a fork and a spoon as manipulators. The manipulators help in eating by griping food. The manipulation is considered effective in some cases. However, owing to the structure of the manipulators, the system has difficulty in griping soft and slippery food. There is also a risk to make fragile food into pieces. Based on the robots introduced above [9, 10], some researches related to meal assistance robots are carried out. It is not difficult to summary the meal assistance robot that the articulated robots introduced are not easy to avoid vibration owing to the structure. Since in taking and passing food procedures, vibration generated on one axis will be transferred to other axis. Another reason why conventional systems are not easy to be applied is that the robots are designed to be with operation interfaces such as joystick which is supposed to be operated with normal hands. In most cases, the operation interfaces are not suitable for upper-limb-crippled people.

Because of the shortcomings introduced above, we propose the meal assistance robot to solve the existent problems with the following features.

- help eating by passing food to user smoothly.
- orthogonal coordinate structure.
- user friendly operation interface with head movement.

2 Meal Assistance Robot

2.1 Structure of Proposed Meal Assistance Robot

Different from previous meal assistance robots our proposed one is structured in an orthogonal coordinates. Figure 1a shows configuration of proposed robot. The main components of the robot are including the pusher which applied to push food from food plate in y-axis direction as shown in the figure; the shutter which can be

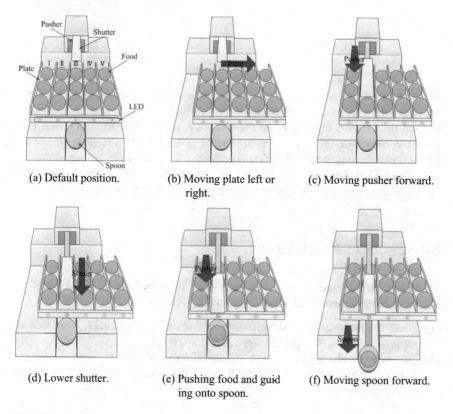

(a) Default position.

(b) Moving plate left or right.

(c) Moving pusher forward.

(d) Lower shutter.

(e) Pushing food and guiding onto spoon.

(f) Moving spoon forward.

Fig. 1 Process for conveying foods

Fig. 2 Flowchart for
proposed system

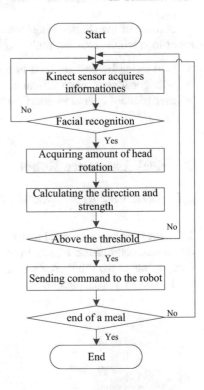

adjusted in vertical direction of z-axis; the food plate with tracks, and the LEDs
which are used for marking the tracks; a spoon which is suitable to carry most kinds
of food to the user. Operation direction of each axis is defined as follows. Plate
moves in only x-axis direction, shutter moves in only z-axis direction, pusher and
spoon move in only y-axis direction. The forces from actuators are transmitted in
axis with rack-and-pinion. The DC motors are employed to the plate, pusher and
spoon as actuators. For the shutter a servo motor is used (Fig. 2).

2.2 Proposed Robot Action

Each axis at proposed robot moves as shown in procedure of Fig. 1 based on
proposed robot structure described at preceding section.

(1) Foods are set up at the predetermined location on the plate tracks. The foods are
    set in five tracks. The pusher stops at the 3rd track.
(2) The user can choose food using the operation interface with head movement.
    The x-axis movement of plate will make the selected track moving to the
    pusher position.

(3) The pusher moves to the position of the selected food.
(4) The shutter comes down to the track.
(5) The pusher pushes out the food to the spoon in standby status.
(6) The spoon extends and passes the food to user's mouth.

The location shutter can be modified according to setting based on different food. The food is supposed to be offered following their locations on the tracks.

2.3   Choice Food and Provide

There are LEDs applied for marking the tracks on the plate. The user can choose the food according to LED on the tracks. In choosing procedure, LEDs sequentially light up. The user chooses the food when LED of the track which contains the objective food lights up. The choice of user will be recognized by operation interface with head movement.

We have developed operable interface which helps people giving information to the computer by head movement. By this way upper-limb-crippled people can operate the proposed robot easily.

2.4   Available Provided Foods

The proposed robot is able to help user with having solid food. Experiment with marketed lunchbox has been implemented to confirm that the proposed robot can provide all of solid foods perfectly on.

Foods from seven kinds of lunchboxes were employed in experiments as subjects. There are 47 kinds of foods in the lunchboxes. The food offering experiments have been repeated for 232 times totally to verify the effectiveness for passing the food introduced above to user. The proposed robot provided all of the foods except four times on non-solid foods.

3   Operation System with Head Movement

3.1   Outline of the Proposed System

The proposed system measures the head motion and sends commands to the robot. Firstly, face detection is done using Kinect sensor by the system. Next, the amount of horizontal rotation and the amount of vertical rotation of the head are acquired. These are expressed in polar coordinate. When the value exceeds a certain threshold value, the proposed system transmutes to the robot. Above process are repeated until all foods are selected.

Kinect sensor is a Natural User Interface developed by Microsoft Corporation. It is a sensor for game machines that can be operated without using a physical controller. Kinect sensor are equipped with RGB camera, depth sensor, multi-array microphone, and tilt motor. It can recognize the position, movement, voice, and face of the user based on the information obtained from these devices.

3.2 Information Processing for Head Movement

In consideration of the use environment, the information acquired by Kinect sensor is limited as follows.

- Set Kinect sensor to Near mode
- One measurement subject
- Target the nearest person from Kinect sensor.

On the above setting, the proposed system measures amount of head rotation in the vertical direction and horizontal direction. Figure 3 show the amount of rotation acquired in each direction. "Pitch" indicates the amount of rotation in the vertical direction, and "Yaw" indicates the amount of rotation in the horizontal direction.

Next, polar coordinate transformation is performed from the obtained head-rotation-amount by expressions (1) and (2). Here, θ is the head inclination viewed from Kinect sensor, r is the head inclination amount seen from Kinect sensor, y_{pitch} is the vertical head-rotation-amount, and x_{yaw} is the horizontal head-rotation-amount. θ and r are expressed in the following equations.

$$\theta = \tan^{-1} \frac{y_{pitch}}{x_{yaw}} \tag{1}$$

Fig. 3 The rotate direction

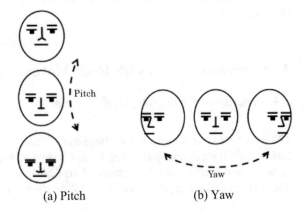

(a) Pitch   (b) Yaw

Fig. 4 Schematic diagram of the information acceptance characteristics depending on the viewing position

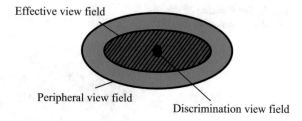

Effective view field

Peripheral view field

Discrimination view field

$$r = \sqrt{x_{yaw}^2 + y_{pitch}^2} \qquad (2)$$

By these equations, it is possible to express the inclination degree of the head viewed from Kinect sensor. When the user selects a food, we refer to the functional distribution of the view field of information acceptance so that the user can eat with natural action [11]. Figure 4 shows a schematic diagram of the information acceptance characteristics depending on the viewing position. Three kinds of visual field functions are described here.

Discrimination view field

The viewing ability is outperform, it is possible to accept information with highly accurate (within about 5° of center)

Effective view field

It is possible to accept information with only head movement (about 15° left and right, about 12° downward, within about 8° upward)

Peripheral view field

By the coordinated movement between the eyeball and the head makes, it possible to watch the information in a stable state. Although the presence of visual information can be determined, it is impossible to discriminate details Information acceptance ability is extremely reduced. It shows an auxiliary function to the extent that a gaze movement is caused against a sudden fluctuation stimulus It is gathered as an area outside the effective view field.

Here, a threshold for dish selection is set using an area where information can be accepted only by eye movement to the effective view field.

3.3 Threshold Setting

Figure 5 shows the situation of using the robot as seen from the side. The distance from Kinect sensor to the eyes is 65 cm. If the user is farther from Kinect sensor than this distance, it becomes difficult to eat because the driving range of the robot

Fig. 5 The situation of using the robot as seen from the side

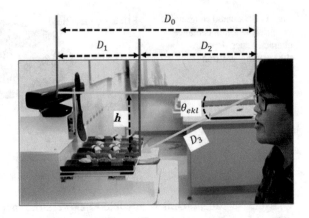

is greatly exceeded. Thus, in this paper, a threshold is set based on the above mentioned values.

First, the head rotation amount in the vertical direction will be described.

For consider the angle θ_{ekl} of the eye, the Kinect and the LED in Fig. 5, the horizontal distance D_1 from the Kinect sensor to the LED is 20 cm, the distance D_2 from the LED to the eye is 45 cm, and the vertical distance h from the LED to the Kinect sensor is 17 cm. With these parameters θ_{ekl} is expressed by the following equation.

$$\theta_{ekl} = \tan^{-1}\frac{h}{D_2} = 20 \quad (\text{deg}) \tag{3}$$

On the other hand, in the effective view field, the view field in the downward direction θ_{evl} is 12°. When the user watch the LED, the inclination of the head in the vertical direction θ_{hl} is expressed by the following equation.

$$\theta_{hl} = \theta_{ekl} - \theta_{evL} = 8 \quad (\text{deg}) \tag{4}$$

As stated above, it is expected that the movement of the head additional the movement of the eyeball rotates downward by 8°, when the user looks at the LED.

Next, the rotation amount of the head portion in the horizontal direction will be described. Figure 6 shows the situation at using robot as seen from the top of the front. The linear distance from the midpoint of both eyes to the center of the plate D_3 is derived by the following equation from the theorem of triple square.

$$D_3 = \sqrt{D_2^2 + h^2} = 48.1 \quad (\text{cm}) \tag{5}$$

The distance from the edge to the center of the dish D_4 is 12.5 cm. The angle from the center of the head to the plate edge θ_{pl} is expressed by the following equation from the inverse trigonometric function.

Fig. 6 The situation at using robot as seen from the *top* of the *front*

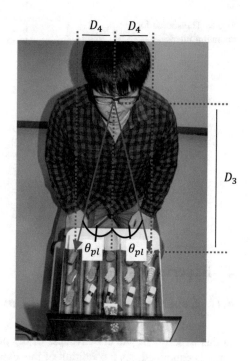

Fig. 7 *Vertical* and *horizontal* relationship as seen at Kinect sensor

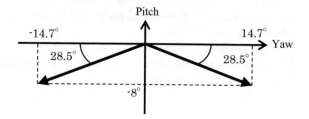

$$\theta_{pl} = \tan^{-1}\frac{D_4}{D_3} = 14.7 \quad (\text{deg}) \tag{6}$$

The relationship between the head rotation amount in the vertical direction and the horizontal direction is summarized in Fig. 7.

Thus, in this paper, the threshold is set as shown in Fig. 8. First, the threshold in the vertical direction is an area below $-8°$. Next, the plates ① and ⑤ on the left and right end are areas exceeding $28.5°$ from the Yaw axis. The plates of ②–④ have been evenly divided at equal intervals of $41°$. Finally, the plate selection is made in the vertical direction by $15°$ or more and the Pitch axis by $\pm 20°$ and $40°$ for preventing malfunction. When this threshold is exceeded, it indicates that plate is selected. Then the system sends an instruction to the robot. As a result, it is possible to manipulate the robot only by movement of the head.

Fig. 8 Thresholds for
operation interface

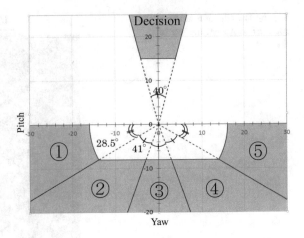

4   Experiments and Results

4.1   Experimental Method and Conditions

In this experiment, the effectiveness of the proposed system in the meal assistance robot is examined. As a condition of this experiment, the position of Kinect sensor was set at a distance of 65 cm from the examinee, and adjusted to the height of the eye. The examinees chosen each plates at three times, and it is success if they could choose the intended plate.

This number corresponds to the number of foods that this robot can assist for one meal. Examinees were 6 men in the 20s: A to F (3 spectacle wearers), and 1 woman in their 20s: G (naked eye).

4.2   Experiment Results and Consideration

The experimental results are summarized in Table 1. Overall, the examinees were able to select the almost all intended plate.

From the experimental results, they selected the intended plate at a high rate regardless of whether or not wear glasses. However, when selecting a Plate ② with a man wearing eyeglasses and a unaided woman, there were false recognitions. When the examinees select Plate ② and the head was tilted upward for decision, it was slightly within the threshold region of Plate ①. This caused false recognitions for the reason. As a method to solve this problem, it is possible to prevent the chattering by sending a signal to the robot when plate selection persists for a length of time. On the other hand, it has a potential lost operability as a delay occurs in the time until the plate decision. It is a future task to set up with functionality while preventing the chattering.

Table 1 Experimental results

Examinee (grass)	Recognition rate				
	Plate ①	Plate ②	Plate ③	Plate ④	Plate ⑤
A (Present)	3/3	3/3	3/3	3/3	3/3
B (Present)	3/3	2/3	3/3	3/3	3/3
C (Present)	3/3	3/3	3/3	3/3	3/3
D (Absent)	3/3	3/3	3/3	3/3	3/3
E (Absent)	3/3	3/3	3/3	3/3	3/3
F (Absent)	3/3	3/3	3/3	3/3	3/3
G (Absent)	3/3	2/3	3/3	3/3	3/3
Total rate	100%	90%	100%	100%	100%

5 Conclusion

In this paper, a meal assistance robot system operated by operation system with head movement is the proposed. The robot was built up with simple structure and stable motion. Meanwhile, the robot can be manipulated in an easy mean owing to the user friendly operation system with head movement proposed in this research.

In the experiments, a consistency rate of 90% or more was obtained irrespective of whether or not glasses were worn. As the results, the effectiveness of the head operation system has been confirmed. It is necessary to combine prevent chattering which is a cause of false recognitions and operability as future tasks.

References

1. Cabinet Office, Government of Japan: Annual Report on Government Measures for Persons with Disabilities (2015)
2. Department of Health and Welfare for Persons with Disabilities, Social Welfare and War Victims' Relief Bureau, Ministry of Health, Labour and Welfare comprehensive: Survey (National Survey of Disabled Children and Person at Nome) Results on difficulty of live at 2011, pp. 1–7 (2013)
3. Topping, M., Smith, J.: The development of handy1, a robotic system to assist the severely disabled. In: Sixth International Conference on Rehabilitation Robotics, pp. 244–249 (1999)
4. Topping, M.: Handy1, A robotic aid to independence for severely disabled people. Integration of Assistive Technology in the Information Age, pp. 142–147 (2001)
5. Topping, M.: An overview of the development of handy1, a rehabilitation robot to assist the severely disabled. J. Intell. Robot. Syst. **34**, 253–263 (2002)
6. Sijs, J., Liefhebber, F., Romer, G.: Combined position & force control for a robotic manipulator. In: IEEE 10th International Conference on Rehabilitation Robotics, pp. 106–111 (2007)
7. Ishii, S.: Meal-assistance robot "my spoon". J. Robot. Soc. Jpn. **21**(4), 44–47 (2003)
8. Ishii, S., Arai, B.: Robot development in the field of welfare. J. Robot. Soc. Jpn. **24**(3), 304–307 (2006)

9. Uehara, H., Higa, H., Soken, T., Namihira, Y.: Trial development of a mobile feeding assistive robotic ann for people with physical disabilities of the extremities. IEEJ. Trans. **131** (10), 1752–1759 (2011)
10. Arai, K., Mardiyanto, R.: Robot arm control and having meal aid system with eye based human-computer interaction (HCI). Trans. Inst. Electr. Eng. Jpn. C **132**(3), 416–423 (2012)
11. Adachi, K., Hamada, T., Nakano, T., Yamamoto, S.: Blink measurement to detect a driver's drowsy state by moving image processing. Trans. Inst. Electr. Eng. Jpn. C **124**(3), 776–783 (2004)

Decision Tree Analysis in Game Informatics

Masato Konishi, Seiya Okubo, Tetsuro Nishino
and Mitsuo Wakatsuki

Abstract Computer Daihinmin involves playing Daihinmin, a popular card game in Japan, by using a player program. Because strong player programs of Computer Daihinmin use machine-learning techniques, such as the Monte Carlo method, predicting the program's behavior is difficult. In this study, we extract the features of the player program through decision tree analysis. The features of programs are extracted by generating decision trees based on three types of viewpoints. To show the validity of our method, computer experiments were conducted. We applied our method to three programs with relatively obvious behaviors, and we confirmed that the extracted features were correct by observing real behaviors of the programs.

Keywords Daihinmin · Machine-learning techniques · Decision tree analysis

1 Introduction

Because results of game informatics research are expected to be applied to fields such as economics and psychology, many games have been studied. Among others, Daihinmin, a popular card game in Japan, is known as a multiplayer imperfect information game. Computer Daihinmin involves playing Daihinmin by using a player program. As a competition for Daihinmin computer programs, UEC Computer Daihinmin Competition (UECda) [1] is annually held since 2006 at The University of Electro-Communications (UEC), and every year, the champion program is getting increasingly stronger [2]. Strong player programs that participated in UECda use machine learning techniques such as the Monte Carlo method. However, predicting the behavior of a program using machine learning is generally

M. Konishi (✉) · T. Nishino · M. Wakatsuki
Graduate School of Informatics and Engineering, The University of Electro-Communications,
Chofu, Japan
e-mail: j.konishi@uec.ac.jp

S. Okubo
School of Management and Information, University of Shizuoka, Shizuoka, Japan

© Springer International Publishing AG 2018
R. Lee (ed.), *Applied Computing & Information Technology*,
Studies in Computational Intelligence 727, DOI 10.1007/978-3-319-64051-8_2

known to be difficult [3]. Hence, studies on the classification of player programs using cluster analysis have been conducted. However, the extraction of the features determining the behavior of each player program is difficult.

This study is intended at extracting features that determine the behavior of various programs used for playing Daihinmin. For this, we propose a method for extracting the features of player programs by collecting the log of the game and performing data mining using decision tree analysis on the log. In our method, a decision tree is generated focusing on three types of variables, the processing time of a turn, situation of specific cards, and type of submission on the empty field. Then, to verify whether the method accurately extracts the features of the program, we analyzed a representative player program (Default, Nakanaka, Snowl) whose behavior is difficult to predict. Results showed that the extracted features can approximately predict the actual behavior.

2   Rule of Computer DAIHINMIN

In this study, we adopt the rules used for UECda-2015. The rules are almost the same as those in UECda-2007, and their details are as follows.

2.1   Basic Rule

For Daihinmin, the game is played by five players. Daihinmin uses 53 cards, which consist of 13 (A–K) Hearts, Clubs, Spades, and Diamonds, as well as a Joker. When beginning a round of the game, each player is dealt 10 or 11 cards. The card strength order is 2, A, K, Q, J, 10, 9, 8, 7, 6, 5, 4, 3. The cards that a player has in the hand are called his "hand", and the player submits cards from his hand during his turn. A round of the game ends when four players win; that is, they eliminate all of their cards (this is called "Agari"), and the titles are provided in the order of the winning players against the loser. The highest title is Daifugo (the grand million-aire), followed by Fugo (the millionaire), Heimin (the commoner), Hinmin (the needy), and Daihinmin (the extremely needy), in that order. After all the players' titles are determined, the cards are exchanged at the beginning of the next round in the following manner. Two cards are exchanged between the Daihinmin and Daifugo, and one card is exchanged between the Hinmin and Fugo. At this time, the Daihinmin and Hinmin must hand over their strongest cards in the hand.

2.2   Submission of Cards

There are three types in which the players can submit their cards: single, group, sequence. Single implies submitting only one card. Group implies submitting multiple cards of the same number simultaneously. Sequence implies submitting three or more cards of the same mark that are consecutive numbers. In addition, a player can pass, which means not submitting cards at his turn. The submitted cards remain in the field until "Nagare", which means clearing the cards from the field. The main conditions of "Nagare" include the case of "8-ender" and that when all players "pass" in the game. If there are no cards in the field (this situation is called a "lead"), the next player can submit any card. If there are some cards in the field (this situation is called a "follow"), the next player can submit only the cards that are of the same type and are stronger than the cards remaining in the field.

There are special situations of the field such as "lock" and "revolution", wherein "lock" happens when cards with the same mark as the last submitted cards are submitted, and only the cards with the same mark can be submitted until "Nagare", and "revolution" happens when a player submits a group of four or more cards, or a sequence with five or more cards; the strengths of all cards are reversed until the end of the round or until another revolution occurs.

There are some special cards in Daihinmin. Main special cards are Joker and the cards with the number 8. If submitted cards contain an 8-card, the "8-ender" occurs, which forces "Nagare" to occur. Joker can be used as the strongest card when using with "single" and can be substituted as any number card when using with a group or sequence.

3   Decision Tree

A decision tree is a data mining technique that generates a conceptual representation of a tree structure in which each internal node represents an explanation for classification and each leaf represents a classification result. The algorithm for generating a decision tree classifies the data of the analysis source so that there is no variation. Variables that are the basis of variation are called input variables, and variables to be classified are called target variables. The generated tree is called a decision tree if the target variable is a categorical variable, and it is called a regression tree if the target variable is a continuous variable. Decision trees are often used for prediction and classification, and are also used for extracting the rules of the data of analysis source [4]. In other games such as mah-jong and shogi, decision trees are used [5, 6].

When we generate a decision tree, several types of variation criteria can be used. In this study, we use entropy as variation. Entropy is a standard of uncertainty in information theory, and when the data is classified into P_1 and P_2, entropy D is given as follows.

$$D = -P_1 \log P_1 - P_2 \log P_2 \tag{1}$$

In decision tree analysis, entropy is used when generating branches, and the algorithm searches for classifications that result in less entropy.

Creating branches of trees is possible until there are no variations. However, the tree becomes unstable because the branch in the latter half of the tree becomes a more irrational classification. The unstable tree is specialized in the data used for generation, so the tree generally cannot adapt the data. So as to not make the tree unstable, we should prune branches appropriately. There are many decision tree algorithms, and in this study, we use rpart, which is based on Classification And Regression Tree (CART).

The CART algorithm prunes the tree as follows. First, CART makes the tree in which variation is the smallest. Then, because the branches in the upper layer of the tree are stable, the algorithm chooses many trees containing branches that are stable as the candidates. By adapting the candidate trees to the verification data, the algorithm obtains the respective misclassification rates. Finally, according to the rates, the algorithm selects the best tree.

4   Proposed Method

In the proposed method, we collect logs of programs, and then we perform the decision tree analysis on the logs. After that, we confirm the extracted features of the program that are correct by comparison with real features of the program. In this section, we will explain the method of collecting logs and generating decision trees.

4.1   Collecting Logs

To take the data necessary for the decision tree analysis, logs of the programs to be analyzed are collected from Computer Daihinmin. At this time, the logs are collected with elements corresponding to the target variables and the input variables of decision tree analysis as elements. The elements of the logs are shown in Table 1.

4.2   Generating Decision Trees

We define the viewpoint of the analysis and generate the decision trees by using elements suitable for that viewpoint. Therefore, extracted features, input variables, and target variables are different based on the viewpoints. In this paper, we propose a decision tree method based on the following three viewpoints. The viewpoints and the element selections were decided by the preliminary experiments.

Analysis 1: Analysis based on processing time of a turn

Each strategy of a program affects its processing time. In fact, the time allocation is very important in shogi, and there is a method to presume the behavior of a system

Table 1 Details of analytic elements

Name	Detail
Number of game	Total number of games
Number of turn	The number of steps since the beginning of the game
Number of "Nagare"	The number of steps since "Nagare"
Number of total cards	Total number of remaining cards in all hands
Number of my cards	Number of remaining cards in my hand
Lead or follow	The situation of the field when submitting cards
Type of submission	There are five basic types; single, group, sequence, pass, follow (when there are cards in the field). In addition, there are five special types that include 8 cards; 8+ single, 8+ group, 8+ sequence, and 8+ follow
Have a Joker	Whether or not I have a Joker in my hand
Have a card with number 8	Whether or not I have a card with number 8 in my hand
Have a group	Whether or not I have a group in my hand
Have a sequence	Whether or not I have a sequence in my hand
Processing time of a turn	The time interval between the server communicating at the start of a turn to the client and receiving the submission card from the client
Player name	The name of the player that acted

by the processing time. In UECda, there are two classes based on the processing time. Therefore, the processing time is an important feature of a program. A regression tree is generated with the log of one program.

The details of the target and input variables are as follows.

- Input variables
  Number of game, Number of turn, Number of "Nagare", Number of total cards, Number of my cards, Have a Joker, Have a card with number 8, Lead or follow.
- Target variable
  Processing time of a turn.

Analysis 2: Analysis based on the situation of specific cards

In Computer Daihinmin, the features of programs appear in how to use 8-ender [3]. A card with the number 8 is one of the special cards in Daihinmin and Joker is a special card too. Since the special cards affect the behavior of programs, we consider the properties of Joker and a card with the number 8.

We generate a decision tree focused on the method of using a specific card from the log of the given program. The specific cards used in this analysis are Jokers and 8-cards. The details of the target and input variables are as follows.

- Input variables
  Number of total cards, Number of my cards, Lead or follow, Type of submission.
- Target variable
  Player name.

Analysis 3: Analysis based on the type of submission at the time of lead

There are three types of submission (i.e. single, group, sequence) and each program has its own way to submit cards. Thus, we focus on the types of submission.

In addition, there are two types of the field (i.e. lead and follow). When the field is follow, submitted cards are affected by the cards on the field, so that the features of the program about the type of submission do not appear. On the other hand, when the field is lead, submitted cards are no affected by the cards on the field, and the program can submit cards freely, so that the features of the program about the type of submission do appear. Therefore, we consider the type of submission when the field is lead.

We focus on logs of the given program and generate decision trees. The details of the target and input variables are as follows.

- Input variables
  Number of turn, Number of total cards, Number of my cards, Have a Joker, Have a group, Have a sequence.
- Target variable
  Type of submissions (single, group, sequence).

5 Computer Experiments

Through computer experiments, the validity of the proposed method was evaluated. Specifically, we collected logs of three player programs whose behavior is relatively clear, and generated each decision tree. Next, we verified whether the extracted features match the behavior of the program.

The collection of logs and generation of decision trees in this experiment were carried out as follows. We added a program to output logs that record the contents of the game to the server program. We played the same player program and collected logs of 1000 games. To generate the decision tree, we used rpart of R, a programming language, for graphically representing statistical analysis and analysis results [7, 8]. In this study, the decision tree was generated by entropy in R. We used two kinds of pruning, the basic pruning of R and pruning by comparing the tree that generated a new decision tree from another log of the same condition. In pruning of R, 10-fold cross validation strategy was performed. We used R partykit to draw decision trees [9].

5.1 Analyzed Program

5.1.1 Default

It is a program that only performs standard operations, and the algorithm is simple. The Default program was used in UECda-2015.

- Lead
  The Default program always submits as many cards as possible, and then prioritizes in the order of sequence, group and single cards.
- Follow
  If the card in the field is single, the program always avoids cards that comprise a group or sequence, and only the weakest card is submitted. The Joker, however, is an exception; it is never avoided and always submitted. If the card in the field belongs to a group or sequence, the program searches for a probable submittable cards without the Joker, and if a suitable submittable cards is not found, the program searches for a probable submittable cards with the Joker. If, however, a suitable submittable cards is found, the weakest type card is submitted.

5.1.2 Nakanaka

Nakanaka, a player program that participated in UECda-2011 and developped by Prof. Satoru Fujita, was a lightweight standard program participating in UECda-2015. Nakanaka is a heuristic program that considers human strategy when playing Daihinmin and combines 23 strategies. The main features of this program are as follows.

- In end of the game, the Nakanaka program searches for "the winning moves" (moves that definitely win the game) up to three turns in advance.
- Submit cards so that the average of the strength of the hand raise.
- Prioritize lock and sequence.

5.1.3 Snowl

Snowl, a winning program at UECda-2010 and made by Mr. Fumiya Suto, was an indiscriminate standard program at UECda-2015. Snowl is a typical program using the Monte Carlo method of "Computer Daihinmin."

The Snowl algorithm searches for the winning moves. If the winning moves is found by searching, Snowl submits the cards of winning moves. If the winning move is not found by searching, Snowl submits cards by Monte Carlo method [10]. The action using the Monte Carlo method is performed by conducting a plurality of simulations, until the end of the game, against each possible action and then selecting the best action. In addition, Snowl learns by weighing the probability of hand allocation in the simulation to estimate the opponent's hand. Snowl performs the simulation in most cases, so it is difficult to predict the concrete behavior of the Snowl from the program source code. However, since it has been a subject of many studies [3], its behavior has now become clear and, thus, predictable.

5.2 Experimental Results

The regression trees generated by analysis 1 are shown in Figs. 1 and 2. Figure 1 represents Nakanaka, while Fig. 2 represents the Snowl algorithm. It must be noted here that we were unable generate Default regression tree in analysis 1.

The decision trees generated by analysis 2 are shown in Figs. 3 and 4. Figure 3 is the decision tree depicting how the Joker is used, Fig. 4 is the decision tree of the card with the number 8. The three bars in each leaf represent "Default," "Nakanaka," and "snowl" from left to right.

The decision trees generated by analysis 3 are shown in Figs. 5, 6 and 7. Figure 5 represents the Default case, while Figs. 6 and 7, respectively, represent Nakanaka and Snowl. The three bars in each leaf represent "group," "single," and "sequence" from left to right.

From these decision trees, the following features can be noticed.

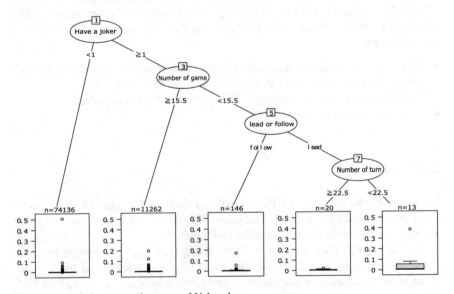

Fig. 1 Analysis 1: a regression trees of Nakanaka

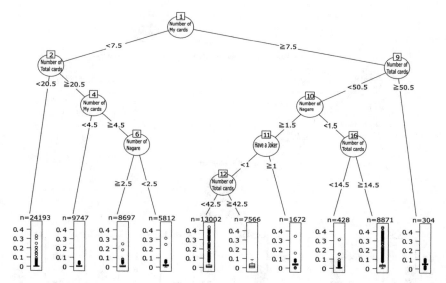

Fig. 2 Analysis 1: a regression trees of Snowl

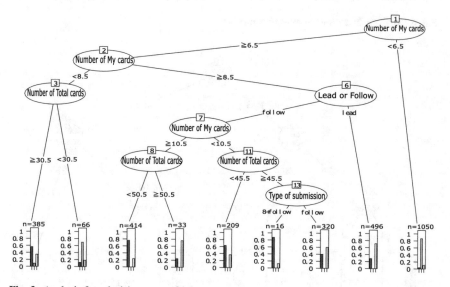

Fig. 3 Analysis 2: a decision trees of joker

5.3 Analysis 1

The fact that the Default decision tree was not generated indicates that the value of the target variable is constant regardless of the input variables. In other words, the Default does not change the processing time for one turn under any situation. From

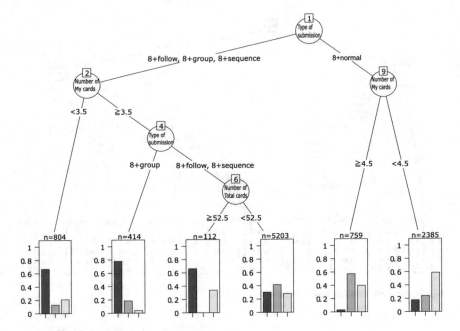

Fig. 4 Analysis 2: a decision trees of 8 number

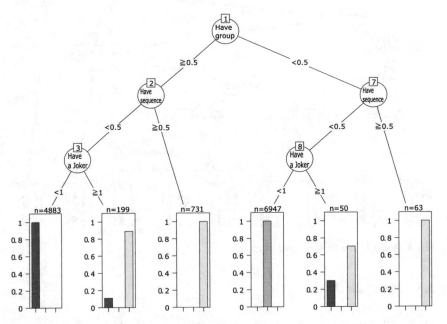

Fig. 5 Analysis 3: a decision trees of Default

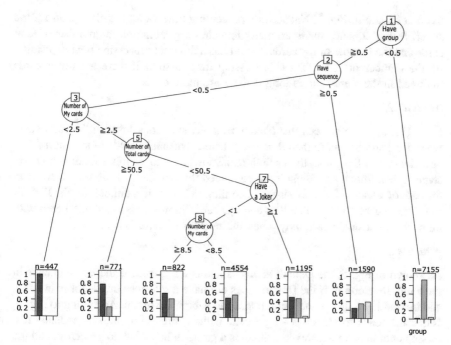

Fig. 6 Analysis 3: a decision trees of Nakanaka

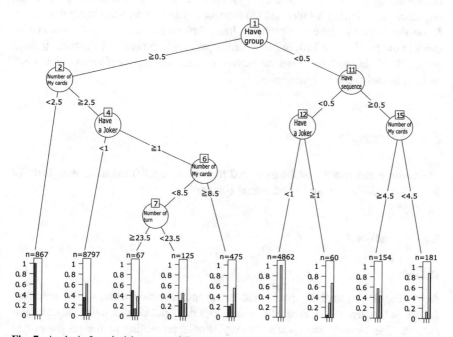

Fig. 7 Analysis 3: a decision trees of Snowl

the first branch in Fig. 1, Nakanaka's processing time varies mainly because if the Joker. In Fig. 2, since there are many branches depending on the number of total cards and the number of my cards, Snowl has a different processing time depending on the number of cards. Thus, processing time is short if there are fewer cards involved in the game, and is long if there are more cards.

Analysis 2

From Fig. 3, it can be seen that Nakanaka never submits a Joker in the early stage when the hand has more than 9 cards. Whereas Nakanaka tends to submit a Joker near the end of the game, the Default and Snowl tend to submit a Joker at an early stage. When Snowl and Default are compared, Snowl tends to submit the Joker in the case of a lead in the early stage. According to Fig. 4, it is difficult for the Default to submit a card with number 8 as a single card. Snowl tends to submit a card with number 8 as a single card only when the number of cards is 3 or less.

Analysis 3

As shown in Fig. 5, the Default always submits a sequence if there is sequence in the hand. In addition, if the Default does not have a sequence or the Joker in hand, and instead has a group, the Default tends to submit the group. According to Fig. 6, Nakanaka tends to submit a single card without branching with a sequence if a group is not in hand. However, if there is a group in hand, but no sequence, and the number of cards in hand is 2 or less, Nakanaka submits the group. Figure 7 demonstrates that even if there is a group in hand, Snowl tends to submit the sequence when having a Joker and the number of cards in the hand being 9 or more. If, on the contrary, there is a group in hand, but no sequence, and the number of cards in the hand is 2 or less, Snowl submits the group, similar to Nakanaka. If there are no groups in hand, but there are sequences, and the number of cards in hand is 4 or less, Snowl submits a sequence.

5.4   Discussion

We examine the validity of the extracted features by comparing them with features found from the source code and actual behavior.

5.4.1   Analysis 1

Default

According to the analysis performed using the proposed method, the extracted feature is the processing time of one turn and does not change under any circumstances. The Default only performs very simple processing in the program, and almost no processing time is required for any occasion. Therefore, the extracted features are valid.

Nakanaka

From the above analysis, it can be inferred that for the Nakanaka algorithm, branching is mainly performed based on the Joker. The Joker can be used as any card when submitted with a group or sequence. Nakanaka has many strategies by avoiding submission of the Joker until the very end of the game. However, when considering winning moves, Nakanaka strategizes the use of the Joker; thus, the influence of the Joker is considered substantial. Therefore, the extracted features are valid.

Snowl

According to the analysis performed using the proposed method, it appears that there are many branches with "number of total cards" and "number of my cards." A lesser number of cards in the hand implies a shorter processing time, more cards imply a longer processing time. Snowl performs a certain number of simulations on actions that can be taken during the game, and as the game goes on, the simulation time goes on reducing. Since the progress of the game is related to the number of cards, this feature has been extracted.

5.4.2 Analysis 2

Default

The Default tends to submit the Joker in the early stage, and it is difficult to submit a card with number 8 as a single card. If the situation of the field is follow and there are submittable cards in the hand, the algorithm always submits cards. Therefore, if the card in the field is a single card, the Default submits a Joker at the end of the game. When the field status is lead, since the Default tends to submit many numbers in the order of sequences, pairs, and single cards, the Default submits cards with the Joker. It is difficult to submit a card with number 8 as a single card because of the low priority assigned to a single. Therefore, the extracted features are valid.

Nakanaka

As observed from the above analysis, Nakanaka abstains from submitting the Joker in the early stage, rather submits it at the end of the game. Nakanaka's algorithm is designed to not submit the Joker until the end of the game, but there is a possibility of submitting the Joker when the field status is revolution. The condition wherein Nakanaka causes revolution is that the strength of the hand increases after submission. Therefore, in the early stage wherein hand consists of many strong cards, Nakanaka does not impose a revolution because it does not lower the average of the hand. Therefore, the extracted features are valid.

Snowl

As observed from the above analysis, Snowl tends to submit the Joker in the early stage and tends to submit the card with number 8 when the number of cards in the hand is 4 or less. Because Snowl is a program based on the Monte Carlo method, the behavior is not explicitly written in the source code. However, observing the behavior of Snowl, it can be seen that the Joker is submitted in the early stage. Therefore, the extracted features are valid.

5.4.3   Analysis 3

Default

As observed from the above analysis, the branch depends on presence of groups and sequences when the field status is lead. This explains the main submission decision process of the Default. In addition, there is feature branching with the Joker. This is thought to be due to the expansion of the choices due to the Joker, which can be substituted for any card, with a group, or sequence. Therefore, the extracted features are valid.

Nakanaka

The branch depends on the presence of groups and sequences. However, in the branch of a group, it tends to submit a single card without branching depending on the presence of the sequence. We presumed that Nakanaka tends to submit a sequence more often than group.

In addition, if there is a group, not sequence and two or less cards in hand, Nakanaka submits the group. The fact that these features appeared in the tree is presumed to be searching for the winning moves at the end of the game. Therefore, the extracted features are valid.

Snowl

As observed from the above analysis, if there are groups in hand then there is no branch about sequence in the tree. In addition, if there is a group, Joker, and the number of cards in the hand of Snowl is 9 or more, Snowl tends to submit a sequence. This condition is the early stage, and Snowl tends to submit a sequence if there is a Joker in the early stage. In addition, there are extracted features such that if there is a group, not sequence, and two or less cards in hand, Snowl submits a group like Nakanaka, and if there is a sequence, not group, and four or less cards in hand, Snowl submits a sequence. These are assumed to be winning moves. Therefore, the extracted features are valid.

6 Conclusion

By generating various decision trees, we extracted many features of Computer Daihinmin programs. For example, Snowl tends to submit a card number 8 in an initial phase. Our conjecture is as follows: the basic strategy of Snowl is to play with a strong card first and then submit weak cards. A card number 8 does not make it difficult to finish the game even when it remains in the card deck, so if Snowl has some strong hands, it tends to submit middle ranking (9 or 10) cards rather than 8. As a result, 8 will remain until the end of the game. However, as the basic 6 strategy of Default is to submit weak cards, 8 will not remain until the end of the game. There are many differences in the method of using the 8-card. By generating a detailed decision tree on the 8-card, we can expect to extract a more detailed strategy.

The analysis of decision trees for Daihinmin, many features can be extracted. This is because that decision tree analysis is strong for missing values, and the model has high interpretability. For Daihinmin, clearly defining the initial, middle, and final phase is difficult. However, in our decision trees, there are many branches asking My cards are X or more. We believe that the number of branches that appeared in the decision tree correspond to initial, middle, and final phases. Moreover, as the branches of trees are not dependent on the Number of total cards but the Number of my cards, the strategy in Daihinmin may strongly be affected by the situation of the individual rather than the overall situation.

References

1. The University of Electro-Communications. UEC Computer Daihinmin Convention (UECda). http://uecda.nishino-lab.jp/
2. Wakatsuki, M., Fujimura, M., Nishino, T.: A decision making method based on society of mind theory in multi-player imperfect information games. Int. J. Softw. Innov. (IJSI) **4**(2), 58–70
3. Ayabe, K., Okubo, S., Nishino, T.: Cluster analysis using N-gram statistics for daihinmin programs and performance evaluations. Int. J. Softw. Innov. (IJSI), **4**(2), 33–57
4. Berry, M.J.A., Linoff, G.S.: Data Mining Techniques: For Marketing, Sales, and Customer Relationship Management (2005)
5. Tanaka, Y., Ikeda, K.: Selection model selection according to circumstances for Mahjong beginners. IPSJ SIG Technical Reports GI, 2014-GI-31(10), 1–8 (2014) (in Japanese)
6. Yanagi, K., Shibawara, K., Tajima, Y., Kotani,Y.: Generation of candidate hands using decision trees in shogi. In: Game Programing Work 2006. pp. 163–166 (2006) (in Japanese)
7. R: The r project for statistical computing. https://www.r-project.org/
8. rpart: Recursive partitioning and regression trees. https://cran.rproject.org/web/packages/rpart/index.html
9. partykit: A toolkit for recursive partytioning. https://cran.rproject.org/web/packages/partykit/index.html
10. Suto, F., Narisawa, K., Shinohara, A.: Development of client snowl for computer daihinmin convention. In: Computer DAIHINMIN Symposium 2010 (2010) (in Japanese)

Binary Blockchain: Solving the Mining Congestion Problem by Dynamically Adjusting the Mining Capacity

Yoohwan Kim and Juyeon Jo

Abstract Mining congestion is a serious issue in blockchain-based cryptocurrencies. It increases the transaction confirmation latency and limits the growth of cryptocurrency. To mitigate the problem, a number of methods have been used in practice and new ideas have been proposed. However, it is not clear whether these schemes can cope with the ever-growing transaction load of cryptocurrencies in the long run. We propose a novel parallel mining method that can adjust the mining capacity based on the transaction load. It does not require an increase in the block size or a reduction of the block confirmation period. In this paper, we describe how and when to divide or merge blockchains, how to solve the imbalanced mining problem, and how to handle the difficulty levels and rewards. We then show the simulation results comparing the performance of Binary blockchain and the traditional blockchain.

Keywords Blockchain · Cryptocurrency · Bitcoin · Binary division

1 Introduction

Many cryptocurrencies, such as Bitcoin, use blockchain to record transactions. To add a block to the blockchain, a mining process is necessary. In the case of Bitcoin, the blocks are generated approximately every 10 min, and each block can be up to 1MB in size. With the transaction size between 250 and 500 bytes, a processing

Y. Kim (✉) · J. Jo
Department of Computer Science, University of Nevada Las Vegas, Las Vegas, NV 89154-4019, USA
e-mail: Yoohwan.Kim@unlv.edu

J. Jo
e-mail: Juyeon.Jo@unlv.edu

© Springer International Publishing AG 2018
R. Lee (ed.), *Applied Computing & Information Technology*,
Studies in Computational Intelligence 727, DOI 10.1007/978-3-319-64051-8_3

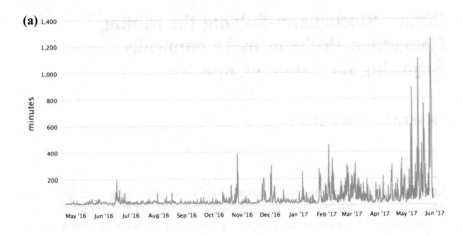

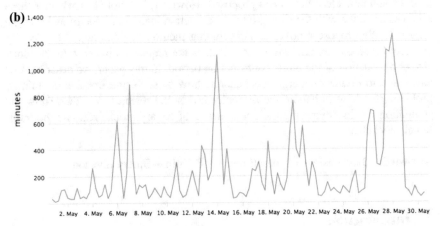

Fig. 1 Average transaction confirmation times [5]. **a** Long term trend over 1 Year (May 2016–May 2017), **b** Short term trend over 1 Month (May 2017)

speed of 3.3–7 transaction/sec can be achieved [1, 2]. Due to increasing popularity of bitcoin, the number of transactions is increasing globally, but the processing speed cannot increase due to design. This limitation creates more backlogs and increases transaction confirmation times, leading to a mining congestion problem [3, 4]. Figure 1a shows the severity of the situation, where the average waiting time for transaction confirmation is growing seriously with the growing popularity of Bitcoin [5]. In May 2016 it was generally below one hour, but in May 2017 it is often exceeding 10 h. Figure 1b shows a daily fluctuation where it ranges between one hour and 20 h.

Table 1 Transaction confirmation times between Feb. 24, 1016 and Mar. 2, 2016

Fees (BTC)	Ave. Confirmation Time (Minutes)	% of transactions with this fee (%)
0	3147	0.84
0.0001	246	7.30
0.0002	55	16.00
0.0003	25	9.07
0.0004	22	42.51
0.0005	20	11.51
0.0006	14	1.38
0.0007	19	1.39
0.0008	13	2.08
0.0009	13	0.59
0.001	12	0.70
(Exceptional cases)		
0.0016	25	0.03
0.007	354	0.00
0.0074	225	0.00
0.0551	1338	0.00

Users can shorten their confirmation time by paying higher transaction fees because miners are more inclined to include the transactions with higher fees in their blocks. Table 1 shows the relationship between transaction fees and the conformation times [6]. Although there are exceptional cases, transactions with higher fees generally get shorter confirmation times.

As a result of mining congestion, many transactions are not confirmed in time and the money is stuck, unable to be used for following transactions. To avoid being stuck in the unconfirmed transaction pool for an extended time, users are forced to pay higher fees, which limits the utility of the blockchain. There are temporary solutions, such as replace-by-fee, child-pays-for-parent, etc. [7], but none directly increase the capacity. Mining congestion is becoming more problematic, as it is limiting the growth of bitcoin and other similar cryptocurrencies that depend on blockchain. The transaction load can increase significantly in certain situations, such as DDoS attacks, which can create an enormous backlog for legitimate transactions. Such congestion may even constrain Bitcoin liquidity and cause prices to fluctuate.

As bitcoin transaction loads go up or down, a new scheme is needed to dynamically adjust the mining capacity based on the transaction load. With that, the transaction conformation speed will be guaranteed and the circulation speed can accelerate. It will make the value of bitcoin more stable, and users can be more confident in the long-term growth of bitcoin.

The issue of blockchain transaction scalability has long been discussed [8–12]. While several methods have been proposed, they are still being debated. We propose a simple method to dynamically adjust the mining capacity based on the transaction load. In the proposed scheme, called Binary blockchain, we multiply the chains when the load goes up; when the load comes down, we reduce the number of chains. Due to the nature of parallel mining and binary division, mining capacity can increase by an order of thousands. In this paper, we describe the process of Binary blockchain management and related issues. We then compare the results of the Binary blockchain simulation with those of the traditional blockchain.

2   Related Work

As the topic of increasing mining capacity is actively discussed in the blockchain community [13], we will review only some of the proposed methods in this section. The current predominant solution is to increase block size [14–17]. While this can temporarily increase mining capacity, block size cannot be increased indefinitely and will eventually face a limit if the bitcoin transactions grow continuously. Another possible solution is to decrease block confirmation time [18]. The downside of this approach is the increased probability of fork and orphaned blocks. Currently, bitcoin block confirmation time is 10 min and forks are created a few times per week on average [19]. Litecoin has already proven the viability of a shorter confirmation time. With the 2.5-minute confirmation time, its probability of having a fork is not very different from Bitcoin's [20]. Although it is conceived that even 12-second confirmation time [21] is possible, this cannot be shortened infinitely as it will cause too much instability to the mining network. Some methods are actually used in the real world, such as off-chain transactions [13, 22, 23], side chain [24], merged mining [25], etc. However, they do not address the capacity of the mining network itself directly. Other innovative solutions have been proposed, such as separating the bitcoin functions on different chains and blocks [26–30]. While these schemes offer a scalable solution, they may depend on other factors such as a larger block size (32MB) to realize enough scalability and may ultimately not catch up with the commercial credit card transaction speed of thousands of transactions per second.

Another approach is to allow multiple branches to confirm the blocks simultaneously. The idea of Tree Chain was proposed in 2014 [31–34], but was only conceptualized and has not progressed enough for further debate. The concept suggests a tree-structured blockchain where each branch can mine blocks, but the structure is static and cannot respond to the dynamically changing transaction load. The use of a DAG (Directed Acyclic Graph) structure instead of a tree structure has also been proposed [35]. MultiChain [36, 37] has an aspect of parallel mining, but it is across different blockchains, not in the same blockchain. Sharding is sometimes mentioned for transaction processing, but is mostly discussed for solving the bitcoin database growth problem [38–41].

Our method, Binary blockchain, is an approach based on parallel mining. In the next section, we discuss the challenges associated with parallel mining in general.

3 Challenges with Parallel Mining

3.1 Double Spending

In a traditional single-chained blockchain, parallel branches or fork may occur inadvertently while the blockchain information propagates throughout the mining network. Since it can create a double spending problem, only one branch can be chosen for further growth. The mining network chooses the longest branch between them, and the orphaned blocks get invalidated.

In parallel mining, multiple branches occur on purpose, not inadvertently, increasing the chances of double spending. To avoid the problem, we have to make sure that each coin is included only in one branch. It is much easier to prevent this from the beginning than invalidating them later. We can divide the transactions into disjoint groups for each branch by taking the hash values of the transactions.

3.2 Decision to Create and Delete New Branches

In parallel mining, there is a new problem of creating a new chain.

1. **Who makes the decision?** If every miner makes the decision, it will be chaotic combined with the network propagation delay. The decision must be unambiguous and universal. The only thing we can trust in the blockchain world is the information contained in the confirmed blocks. So the decision must be made based on the information in the block chain, not arbitrarily.
2. **When do we make the decision?** The decision criteria must be unambiguous to avoid instability. The purpose of parallel mining is to respond to the changing transaction load dynamically. The transaction load can be measured in multiple ways, for example, counting the number of transactions in the mempool [42]. But mempool is not an ideal candidate because it is available only locally. A possible universal measurement of congestion is the size of the confirmed blocks in the blockchain. When there is mining congestion, the size of the block increases. When the block size is close to its maximum size (1MB), we can

assume that there is a backlog [43]. However, miners can arbitrarily make any-sized blocks, so the block size may not measure the congestion level faithfully. A better approach is using the transactions conformation times. When there is congestion, the time for waiting until confirmation for transactions increases, and it results in an increased average confirmation times in the confirmed blocks. This can be observed by all miners and thus used for making a branching decision unanimously.

3.3   Difficulty Level Adjustment

The difficulty level of the mining is adjusted periodically (about every two weeks) to make the average block confirmation time 10 min. Under parallel mining, the miners are divided into multiple groups, and consequently, the conformation time will increase. Therefore, the difficulty level must be reduced accordingly to maintain the 10-minute confirmation time.

3.4   Unbalanced Mining and 51% Attack

When the difficulty level is reduced, these risks exist.

1. **Over-mining on one branch**: Intentionally or unintentionally, all miners may concentrate on one branch. Then the confirmation time for a block will be reduced well below 10 min. This increases the possibility of fork, and eventually more orphaned blocks across the mining network, and cause starvation on the other branch where little mining operation is performed.
2. **Easier 51% attack**: A mega-miner can concentrate on one branch, and launch a 51% attack more easily.

To avoid the risks, there must be a way to ensure balanced mining across all branches. However, there is no miner ID or and miners may choose to min any branch. So a balancing mechanism must be devised.

3.5   Reward Size

With parallel mining, there are multiple branches at the same time. If the reward amount per block stays same, the total reward will increase, which will violate the

Table 2 Definitions

Terms	Definition
Traditional chain	The single original blockchain
Branch	Temporarily competing blockchains in the traditional blockchain
Fork	Process of creating branches (unintentionally)
Main chain	A linear portion of Binary blockchain
Parent chain	A portion of Binary blockchain before division
Child(ren) chain	A portion of Binary blockchain after division
Subchain	Same as child(ren) chain
Sibling chains	A pair of subchains
Division	Creating a pair of chains
Merge	Combining a pair of subchains into one chain

design principle of the current bitcoin system. Moreover, miners will have a great incentive to make parallel branches to win the reward more easily. To prevent it, the size of the reward must be reduced so that the amount of total reward is same as in the traditional blockchain.

4 Proposed Solution: Binary Blockchain

We propose a solution that can dynamically adjust the mining capacity to the overall mining network congestion level. From now, we use the terminology defined in Table 2.

We solve the transaction speed scalability using the following methods.

1. Binary division of block chain
2. Deterministic creation and deletion of the subchains
3. Binary sharding of transaction
4. Binary adjustment of difficulty levels and reward amount
5. Balanced mining with binary synch blocks.

The process of creating and merging the subchains is described in Algorithm 1. We explain the process in detail in the following sections.

Algorithm 1: Block Creation with Division and Merge

Input: T (the set of transactions), C (existing Binary blockchain)
Output: B (new block)

1   // Choose a subchain
2   Choose any subchain from C (or main chain if there is only one chain)

3   // Check for division
4   **If** division conditions are met
5 | Decide which subchain to follow between new subchains
6   **Else**
7 | Follow the current subchain
8   **End**

9   // Check for synch block
10 **If** the block B to be created is a synch block
11 | **If** the hash values from all sibling pre-synch blocks are available
12 | Inherit them all
13 | **Else**
14 | Go to 2 (Choose another subchain from C and re-start)
15 | **End**
16 **Else**
17 | Inherit hash value only from the preceding block
18 **End**

19 //Check for merge
20 **If** merge conditions are met
21 | Inherit hash from both subchains
22 **Else**
23 | Inherit hash value only from the preceding block
24 **End**

25 // Block numbering
26 Assign an appropriate block number to the new block B

27 // Transaction sharding
28 Choose eligible transactions from T for the new block B

29 // Perform mining
30 Confirm block B, and add it to the chosen subchain if confirmed

---

4.1 Binary Division

It is possible to increase the number of chains linearly one by one, (1, 2, 3,…), but it is hard to manage in terms of transaction grouping, and mining load balancing. When the number of subchains increase or decrease in power of 2, it is much easier to manage. Therefore, we propose a binary division of the block chain and call it Binary blockchain.

A binary division of the blockchain creates a pair of subchains and the total mining capacity increases by twice. Each subchain can be divided further or merged independently based on its own transaction load. When it divides, both new blocks inherit the hash value from the parent block, thus maintaining the blockchain. On each subchain, the blocks inherit hash values only from the immediately preceding blocks, except synch blocks. With the proposed Binary blockchain, we have these advantages.

- **Subchain creation and deletion process**: The creation and deletion processes are simple binary operation, i.e., dividing the current branch into two or merging two into one.
- **Transaction grouping**: The transactions can be divided into disjoint groups for each subchain based on their hash values with simple modulo operation by the level of subchain. This removes the possibility of double inclusion in multiple subchains.
- **Difficulty Level**: The difficulty level gets halved for each division. This increases the overall mining capacity by a factor of two.
- **Reward calculation**: The reward gets halved for each division. This ensures that the amount of total rewards stays same as in the traditional blockchain regardless of the number of subchains.
- **Continued growth**: Theoretically the division can be repeated up to the number of hash bits.
- **Balanced mining**: Balanced mining can be maintained systematically with the synch blocks. The distance between synch blocks increase by power of 2 after each division.

Although it may look similar, Binary blockchain is different from TreeChain [33, 34] in that the whole blockchain or each subchain can dynamically and independently increase or decrease. It is also different from side chain [33, 34] or data sharding.

4.2 Block Numbering Scheme

Since the blockchain is not linear any more, we need to number the blocks differently. We use a hierarchical numbering format of "n1.n2.n3...", where a division is marked by a period symbol (".") and each number indicates the location within the subchain. The block number increases only within the last subchain level. In Binary blockchain, two subchains (top and bottom) are created upon division, and we need to differentiate them. For that, we divide the numbers in two groups (odd and even) and assign them to each subchain. In the top subchain, the block numbers grow in even numbers (0, 2, 4, ...), and in the bottom subchain, they grow in odd numbers (1, 3, 5, ...). When the subchains are merged, the last level subchain block number is removed and the upper level subchain numbering continues. An example is shown in Fig. 2.

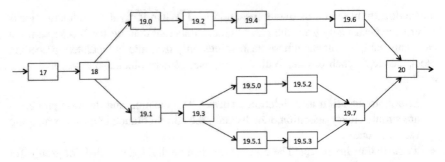

Fig. 2 Block numbering example in binary blockchain

4.3 Balanced Mining with Synch Blocks

Without some kind of synchronization among the subchains, one subchain can progress rapidly, while the other one starves. Such a rapid progress on one subchain creates instability in the P2P mining network because a block gets confirmed much faster than 10 min. It may also create a biased rewarding toward large miners.

One solution is forcing all sibling blocks to inherit the hash values from the other subchain. While it ensures that every subchain progresses at the same pace, however, it causes some overhead of inheriting many hash values. And more importantly, it creates a bigger instability because all miners will move to another subchain as soon as one subchain is mined.

To ensure balanced mining, we introduce a binary synch block. A synch block must inherit the hash values from all pre-synch blocks in the sibling chain. So until all pre-synch blocks are confirmed, no miner can proceed further. The synch blocks are placed as following. Let v be the level of the subchain. The main chain has the level of 0, and the first division creates level 1 subchains, etc. Then at level v, synch blocks are placed every 2^v block. At level v, the difficulty level is 2^{-v} of the original difficulty. Figure 3 shows the example of level 1 subchain, where the synch blocks are placed every 2 blocks. The total amount of work between the synch blocks on Binary blockchain is equivalent to the amount of work for one block in traditional blockchain. Generally,

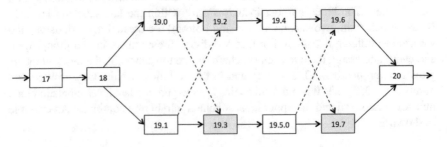

Fig. 3 Synch blocks

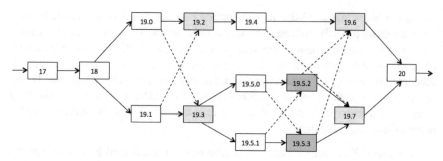

Fig. 4 Global synchronization

- Number of blocks between synch blocks at level $v = 2^v$
- Difficulty level at level $v = 2^{-v}$
- Reward for one block at level $v = 2^{-v}$
- The total amount of work between synch blocks $= 2^v \times 2^{-v} = 1$
- The total amount of reward between synch blocks $= 2^v \times 2^{-v} = 1$

For example, when there are two subchains, the difficulty level is halved, and the number of blocks between the synch blocks is two including. So the amount of work for two blocks is equivalent to the amount of work for one block in traditional chain.

With this scheme, even if all miners are concentrated on one subchain, the maximum number of blocks they can continuously confirm is equivalent to the amount of work in one block in a classical blockchain. For example, if there are 4 sub-branches, each block can take 2.5 min to mine if all miners work on this subchain. These 4 consecutive blocks acts like one regular block. The total amount of reward between synch blocks is also equivalent to the reward for one block in traditional blockchain.

The synchronization must be done at all levels to ensure global balancing for all subchains. For example, in a complete Binary blockchain at level v, the number of subchains is 2^v, and the synch blocks must inherit the hash values from all 2^v sibling blocks. This forces global synchronization across all subchains. Figure 4 shows an example of global scale synchronization.

4.4 Division and Merge Decision

In Binary blockchain, the decision to divide or merge is pre-determined by the blockchain itself, not arbitrarily determined by any individual. As mentioned before, there are ways of measuring the mining congestion level. It is probably most accurate to count the number of transaction in mempool, but it varies widely among miners and it is hard to make a consensus. The information must come directly from the blockchain itself. One possibility is the size of the blocks. Currently the maximum block size is 1MB, but it fluctuates between 500 KB and 1MB. When it becomes close to 1MB, we can assume that there are more transactions. However, a more accurate measurement is the average time for confirmation for the transactions included in the confirmed blocks. When there is mining congestion, there is competition among the transactions and the time for confirmation grows. In this research, we use the average confirmation time to make a decision to divide or merge. For this purpose, the confirmation time doesn't need to be accurate and the order of minutes is sufficient.

(1) **Division**: When the average transaction times in confirmed blocks go above a threshold value (e.g., 20 min), the network divides the chain. To avoid fluctuation, it divides only when the average confirmation times go above threshold over multiple blocks continuously (e.g., 3 blocks).

(2) **Merge**: The subchains are merged when the average transaction times go below a certain threshold (e.g., 10 min) for multiple consecutive blocks. If one of the subchains meets the merge condition, the merge occurs regardless of the transaction load of the other subchain. The merge can occur only at one level at a time.

For simplicity, the merge can occur only at synch blocks. When one of the subchains meets the merge condition at a synch block, it must wait until the other subchain generates a synch block. Then the next block is the merged block and it inherits the hash values from both subchains.

5   Advantages of Binary Blockchain

5.1   Scalability

Currently bitcoin mining network can process about 3 transactions per second. If the number of bitcoin transactions grows to the commercial credit card level (~ 2000 per second), it will require about 700 times more capacity. This is nearly impossible even with the proposed larger blocks sizes. With Binary blockchain, with just 10 divisions ($2^{10} = 1024$ subchains), the capacity of the mining network can be increased to 3000 transactions per second, matching the commercial credit card transaction speed.

5.2   Cost-Effectiveness

Since the overall mining cost does not increase under Binary blockchain, and only the number of processed transaction increases, the cost per transaction becomes lower. This allows a cost-effective growth, which is an advantage compared with the credit card processing systems. The cost per transaction can become even less than credit card transaction cost eventually.

5.3 Attack Resistance to DDoS Attack

Since Binary blockchain can respond to the transaction load dynamically, the number of subchains is increased automatically during a DDoS period [44], and decrease when the attack is finished. Thus it can effectively cope with a DDoS attack. It should be noted that, however, Binary blockchain is designed to move slowly over hours, not in minutes. So if the DDoS attack fluctuates rapidly, e.g., repeating every half hour, it may not respond quickly enough.

5.4 Smaller Mining Pools

Currently because the chance of successful mining is low, small miners prefer to join mining pools and share the reward. With Binary blockchain, the difficulty level is lower and the reward is also lower, which has a similar statistical effect as pool mining. Therefore, without joining a mining pool, small miners can have a better chance of successful mining. This will increase non-pool miners and further decentralize the blockchain operation positively (e.g., less chance of 51% attack).

5.5 Lower Transaction Fee

Under mining congestion, users are forced to pay a higher transaction fee. With the increased mining capacity, transaction with lower or no fees can still be included, which better serves the original intent of cryptocurrency.

6 Risks and Solutions

6.1 False Timestamp Values

A user can create a transaction with any time value, intentionally or accidentally. This does not give any advantage or disadvantage to the user. But when the transaction is included in a confirmed block, it will affect the average transactions times and cause inappropriate divisions or merges. Generally, any time values in a decentralized blockchain network are not trustworthy, even in the confirmed blocks

[45, 46]. How do make sure the time values in the transactions are correct? While this needs further research, we propose 2 quick solutions.

(1) **Community approach**: Create bitcoin wallet applications and mining software to obey the current time. This is easy to implement because we need an accuracy of only in the order of minutes. Also the receivers ask for a reasonably accurate timestamp in the transmitted bitcoin. Furthermore, we can consider only the middle 50% of the transactions for calculating the average transaction times. Then as long as the majority of the users do not fake the time, the impact will be minimal.

(2) **Changed values between blocks**: Instead of using the average transaction times directly, we can use the changes in average transaction times. When the average transaction times increase by a pre-defined value (e.g., 2 min) over multiple blocks, we can decide to divide. Attackers can still fake the time values in a large amount of transactions, but it is hard to manipulate the change values between the blocks. The blocks may be mined anywhere in the world, and the transactions included in the block may be very different. Even if the attacker can increase or decrease the average transaction time in one particular block, the effort will become ineffective in the next block.

6.2 Instability Due to Excessive Forks

In Binary blockchain, like in the traditional blockchain, multiple competing branches may be created inadvertently. Then miners will continue mining and eventually the longer branch will win. When a parent subchain turns out to be an orphan branch, all of the subchains become orphaned together. When the block confirmation time is decreased, the chance of fork also increased due to network propagation delay. Unfortunately, because it is now less difficult to confirm a block in a subchain, it will take much less time to confirm a block if all miners work on one subchain globally. For example, after 4 divisions, there will be 16 subchains and the block confirmation time can be 1/16 of 600 s, which is 38 s. This can cause some instability in the mining network, creating more forks.

First of all, such an extreme concentration is unlikely in a widely distributed world of blockchain. Such an event will be possible only when the miners collaborate to make their own mining network unstable. When there are more orphan blocks, the more time and energy they waste, which is not beneficial to miners. Therefore, miners are more likely to diversify the choice of subchains for their own sake [47].

Secondly, the synch blocks are designed to prevent this kind of bias to some degree. While waiting for a sibling's pre-synch block, miners are forced to confirm other subchains. This forces the miners to divide their computing power. When a branch is forked, the longer (or heavier) chain wins. Whether it is length-based (the length is defined as the slowest-progressing synch block among the sibling subchains) or weight-based, miners must generate as many blocks as possible in their own branch to win over potential competition.

With enough diversification, the generation of each block in any subchain will be still about 10 min. Therefore the chance of fork is not greater than in the traditional blockchain.

7 Performance Evaluation

7.1 Scalability Analysis

In a traditional blockchain, the average mining throughput is given as following.
Mining throughput (= number of transactions/second) = $S_B/S_T/T_B$, where

- S_B = Maximum size of a block (Currently 1MB)
- S_T = Average size of each transaction (250–500 bytes)
- T_B = Average block conformation period in seconds (600 s on average)

With the above typical values, the mining throughput is 3.33. (= 1MB/500 Bytes/600). Multiple proposals are in play to increase the block size. If the block size gets increased, the throughput goes up linearly. For example, with a 4 KB block size, the throughput will be increased to about 13 transactions/sec.

In Binary block chain, the throughput is increased linearly with the number of subchains. For example, with 10 subchains the throughput is increased by 10 times, i.e., 33.3 transactions per second. Note that the number of subchains may not be a power of 2 because each subchain can be split independently.

7.2 Average Confirmation Time Analysis

For most Bitcoin users, the ultimate concern is the confirmation time for their transactions, not the mining network throughput. A normal queuing process does not apply because the confirmation is a result of random selection. In particular, the confirmation time follows a geometric probability distribution [48]. The average confirmation time depends on the overall transaction load in the whole Bitcoin mining network. The probability of being included in the next block for a transaction is N_B/N_T. where,

- N_T = Total number of pending transactions in the mining network
- N_B = Number of transactions in each block

Let N_B/N_T be denoted by p_1. If the number of incoming transactions per block confirmation period is always same as N_B, the total number of pending transactions (N_T) is constant. Then the probability to be included in the n-th block (= p_n) for a transaction is,

$$p_n = p_1 * (1 - p_1)^{n-1}$$

In a geometric distribution, the average and variance are given as following.

$$E(n) = \frac{1}{p_1}, \quad var(n) = \frac{(1 - p_1)}{p_1^2}$$

In case of Binary blockchain, the confirmation time is given as follows.

$$p_n = (p_1 * N_c) * (1 - p_1 * N_c)^{n-1}, \quad where$$

- N_c = Number of subchains
- $\lceil p_1 * N_c \rceil = 1$

The average and variance are given as following.

$$E(n) = \frac{1}{p_1 * N_c}, \quad var(n) = \frac{(1 - p_1 * N_c)}{(p_1 * N_c)^2}$$

For a comparison, the confirmation times in both cases are shown in Table 3 and also graphically illustrated in Fig. 5 under the following conditions.

- $N_T = 10,000$
- $N_B = 2000$
- $N_C = 2$
- $p_1 = 0.2 \ (= 2000 / 10,000)$ for traditional blockchain, or
  $p_1 = 0.4 \ (= 2000*2 / 10,000)$ for Binary blockchain

The corresponding statistical values in both cases are given in Table 4.

We can observe that the conformation time gets reduced greatly by increasing the throughput by twice. In the actual blockchain environment, the incoming

Table 3 Probability of confirmation within n-th block period

Block period	Traditional blockchain	Binary blockchain with 2 subchains
1	0.200	0.400
2	0.360	0.640
3	0.488	0.784
4	0.590	0.870
5	0.672	0.922
6	0.738	0.953
7	0.790	0.972
8	0.832	0.983
9	0.866	0.990
10	0.893	0.994

Fig. 5 Probability of confirmation within n-th block period

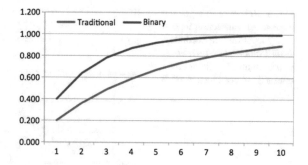

Table 4 Statistical values of confirmation times

	Traditional blockchain	Binary blockchain with 2 subchains
Average	5	2.5
Variance	20	3.75
Standard Deviation	4.47	1.94

transaction load fluctuates significantly, and a more advanced analysis will be needed. We leave it for future studies.

7.3 Simulation Results

To test the scalability of the Binary blockchain, we performed simulation under the following conditions.

- Total number of transactions per block = 500
- Block confirmation period = 10 min (= 600 s)
- Number of block periods = 50 (= 500 min)
- Transaction fee = none (not considered)

We generated the transaction as following. First, we generated 500 transactions per block period with the initial transactions of 1000. In each block, 500 transactions are selected randomly. In this case, there is no backlog and the behavior of the traditional and Binary blockchain was identical. Second, the transactions are generated uniformly at the speed of 1000 transactions per second. Figure 6 shows the results. Since there are 500 transactions not included in a block each period, it creates a gradually increasing mining congestion. The result is an increased average transaction confirmation time due to the growing waiting period. In case of Binary blockchain, when the average confirmation time is over 3000 s continuously for 3 blocks, it divides the blockchain. Reversely, if the average confirmation time is below 1500 s for 3 blocks continuously, it merged the subchains. The traditional blockchain shows gradually increasing average transaction confirmation time. The

Fig. 6 Response to a gradual
increase of transaction load.
(*left* y-axis: time in seconds,
x-axis: block period, *right*
y-axis: number of incoming
transactions per block period)

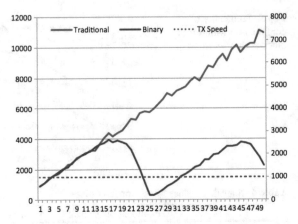

(left y-axis: time in seconds, x-axis: block period
right y-axis: number of incoming transactions per block period)

Binary blockchain shows quick drop in the average confirmation time after a division (around period 17) and consumes most of the backlogged transactions. At the lowest point (period 29), the confirmation time went down to 299 s. Then the subchains are merged, thereby having a more consistent average confirmation time within a range.

Third, we simulated a sudden surge of the transaction load such as in DDoS attack. Figure 7 shows the result. In this case, there was a large amount of incoming transactions (up to 4000 transactions per block period) between the block periods 19 and 25. As expected, the traditional blockchain couldn't handle the transaction load and the average confirmation time kept increasing. In Binary blockchain, the blockchain was divided first when the normal overload was observed (around period 17). Then when the surge hit, it divided again (around period 25) and reduced the confirmation time down to 637 s (period 37). Then it merged as the

Fig. 7 Response to a
temporary surge of
transaction load

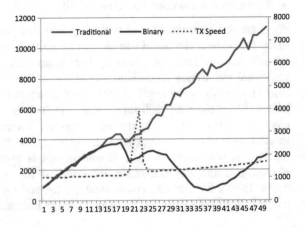

load subsided average and the transaction time went up to the normal range. This results show that Binary blockchain can adjust the capacity to the changing load effectively.

8 Conclusions

Mining congestion is a serious problem that limits the growth a blockchain-based cryptocurrency. Although many schemes have been proposed to resolve the issue, it is not clear yet if they can scale up to the level of commercial credit card transaction processing speed. In this research, we have proposed a dynamically scalable solution called Binary blockchain. It takes advantage of the simplicity of binary operation on division, merge, difficulty level adjustment, and reward adjustment. To prevent imbalanced mining, it employs a synch block system. The decision to divide or merge is made by the blockchain itself, so every miners can follow the decision unanimously. Binary blockchain can adjust the mining capacity according to the transaction load, thereby providing a more consistent confirmation time regardless of the load. We have tested its performance in simulation and observed that Binary blockchains successfully adjusts the mining speed according to the transaction load. Although the actual parameters, such as the threshold times or the number of consecutive blocks, should be further studied, the experiment demonstrates the validity of the Binary blockchain concept. We plan to improve the simulation model and experiment with more scenarios in the future.

References

1. Tschorsch, F., Scheuermann, B.: Bitcoin and Beyond: A Technical Survey on Decentralized Digital Currencies. IEEE Commun. Surv. Tutorials **18**(3), 2084–2123. third quarter (2016)
2. Croman, K., Decker, C., Eyal, I., Gencer, A. E., Juels, A., Kosba, A., Miller, A., Saxena, P., Shi, E., Sirer, E. G., Song, D. and Wattenhofer, R.: On scaling decentralized blockchains. In: 3rd Workshop on Bitcoin Research (BITCOIN), Barbados, February (2016)
3. Tom, S.: Technical Roadblock Might Shatter Bitcoin Dreams. MIT Technology Review, 16 Feb 2016. https://www.technologyreview.com/s/600781/technical-roadblock-might-shatter-bitcoin-dreams/
4. Gilbert, D.: Blockchain Complaints Hit Record Level As Bitcoin Transaction Times Grow And Fees Rise. 8 March 2016. http://www.ibtimes.com/blockchain-complaints-hit-record-level-bitcoin-transaction-times-grow-fees-rise-2332196
5. Average Confirmation Time, https://blockchain.info/charts/avg-confirmation-time. Accessed on 31 May 2017
6. Average confirmation times, https://www.reddit.com/r/Bitcoin/comments/48m9xq/average_confirmation_times/. Accessed on 31 May 2017
7. Breaking Through Congestion, https://medium.com/@alpalpalp/breaking-through-congestion-a60b6d1c9d3#.9kb383578. Accessed on 31 May 2017
8. Scalability FAQ, bitcoin wiki, https://en.bitcoin.it/wiki/Scalability_FAQ. Accessed on 31 May 2017

9. Chernyakhovsky, A.: Bitcoin Scalability: An Outside Perspective. 28 September 2015. https://medium.com/mit-media-lab-digital-currency-initiative/bitcoin-scalability-an-outside-perspective-dd7fde962220#.1i44lqxas
10. James-Lubin, K.: Blockchain scalability: A look at the stumbling blocks to blockchain scalability and some high-level technical solutions. 22 January 2015. https://www.oreilly.com/ideas/blockchain-scalability
11. McConaghy, T.: Blockchain Scalability Part I—The Problem. 14 February 2015. http://trent.st/blog/2015/2/14/blockchain-scalability-part-i-the-problem.html
12. Vukolic, M.: The Quest For Scalable Blockchain Fabric: Proof-of-Work vs. BFT Replication. In: Open Problems in Network Security—IFIP WG 11.4 International Workshop, iNetSec 2015, pp. 112–125. Zurich, Switzerland, October 2015
13. Torpey, K.: 6 Possible Solutions for Bitcoin Scalability. 30 June 2015. https://www.coingecko.com/buzz/six-possible-solutions-for-bitcoin-scalability
14. Quentson, A.: Cornell Study Recommends 4MB Blocksize for Bitcoin. 31 March 2016. https://www.cryptocoinsnews.com/cornell-study-recommends-4mb-blocksize-bitcoin
15. Rizzo, P.: Making Sense of Bitcoin's Divisive Block Size Debate. CoinDesk, 19 Jan 2016. http://www.coindesk.com/making-sense-block-size-debate-bitcoin/
16. Bitcoin Block Size Debate Survey. September 2015. http://bravenewcoin.com/assets/Blockchain-Scalability-Survey-2015/BNC-The-Blockchain-Scalability-Survey-2015.pdf
17. Block size limit controversy, biotcoin wiki. https://en.bitcoin.it/wiki/Block_size_limit_controversy, Accessed on 31 May 2017
18. Hertig, A.: A Lower Block Time Could Help Bitcoin Scale, But Will It Work? CoinDesk, 26 Oct 2016. http://www.coindesk.com/lower-bitcoin-block-time-scale/
19. Number of Orphaned Blocks. https://blockchain.info/charts/n-orphaned-blocks. Accessed on 31 May 2017
20. Litecoin Blockchain Explorer. CryptoID, https://chainz.cryptoid.info/ltc/#!extraction. Accessed on 31 May 2017
21. Buterin, V.: Toward a 12-second Block Time. https://blog.ethereum.org/2014/07/11/toward-a-12-second-block-time. Accessed on 31 May 2017
22. Poon, J., Dryja, T.: The Bitcoin Lightning Network: Scalable Off-Chain Instant Payments. Technical Report. https://lightning.network/lightning-network-paper.pdf. Jan 2016, Draft Version 0.5.9.2
23. Sofia: How the Bitcoin Lightning Network Could Solve the Blockchain Scalability Problem. 6 June 2016. https://letstalkpayments.com/how-bitcoin-lightening-network-could-solve-blockchain-scalability-problem
24. van Valkenburgh, P.: What are Forks, Alt-coins, Meta-coins, and Sidechains? 8 Dec 2015. https://coincenter.org/entry/what-are-forks-alt-coins-meta-coins-and-sidechains
25. What is merged mining—Bitcoin & Namecoin—Litecoin & Dogecoin? 25 Oct 2016. https://www.cryptocompare.com/mining/guides/what-is-merged-mining-bitcoin-namecoin-litecoin-dogecoin
26. Eyal, I., Efe Gencer, A., Gün Sirer, E., van Renesse, R.: Bitcoin-NG: A Scalable Blockchain Protocol. In: 13th USENIX Symposium on Networked Systems Design and Implementation, Santa Clara, CA, Mar 2016
27. Eyal, I., Sirer, E. G.: Bitcoin-NG: A Secure, Faster, Better Blockchain. 14 October 2015. http://hackingdistributed.com/2015/10/14/bitcoin-ng
28. Bradley, J.: On Scaling Decentralized Blockchains. Cryptocoins news, 18 Feb 2016. https://www.cryptocoinsnews.com/scaling-decentralized-blockchains
29. Jovanovic, P.: ByzCoin: Securely Scaling Blockchains. 4 August 2016. http://hackingdistributed.com/2016/08/04/byzcoin
30. Kogias, E. K., Jovanovic, P., Gailly, N., Khoffi, I., Gasser, L., Ford, B.: Enhancing Bitcoin Security and Performance with Strong Consistency via Collective Signing. In: 25th USENIX Security Symposium, Austin, TX, August 2016
31. Sidechains, Treechains, the TL; DR: Welcome to join discussion. 2 August 2014. https://bitcointalk.org/index.php?topic=721564.0

32. Todd, P.: Re: [Bitcoin-development] Tree-chains preliminary summary. 22 March 2014. https://www.mail-archive.com/bitcoin-development@lists.sourceforge.net/msg04388.html
33. Infante, A.: Tree Chains vs. Side Chains: The Controversy Explained. 24 October 2014. https://coinreport.net/tree-chains-vs-side-chains-controversy-explained
34. Sanders, G.: Sidechains, Treechains, the TL; DR. 13 June 2014. https://blog.greenaddress.it/2014/06/13/sidechains-treechains-the-tldr
35. Lewenberg, Y., Sompolinsky, Y., Zohar, A.: Inclusive block chain protocols. In Financial Cryptography and Data Security. In: 19th International Conference, FC 2015, pp. 528–547. 2015
36. Greenspan, G.: MultiChain Private Blockchain—White Paper. Coin Sciences Ltd., http://www.multichain.com/download/MultiChain-White-Paper.pdf. Accessed on 31 May 2017
37. Buterin, V.: Scalability, Part 3: On Metacoin History and Multichain. 13 November 2014. https://blog.ethereum.org/2014/11/13/scalability-part-3-metacoin-history-multichain/
38. BUIP024: Extension Blocks with Address Sharding. Bitcoin Forum. https://bitco.in/forum/threads/buip024-extension-blocks-with-address-sharding.1353/. Accessed on 31 May 2017
39. Zamfir, V.: Scaling Bitcoin Workshop: Montreal 2015—Sharding the Blockchain. https://scalingbitcoin.org/transcript/montreal2015/sharding-the-blockchain. Accessed on 31 May 2017
40. Sharding FAQ. Ethereum Wiki. https://github.com/ethereum/wiki/wiki/Sharding-FAQ. Accessed on 31 May 2017
41. EIP 105 (Serenity): Binary sharding plus contract calling semantics. https://github.com/ethereum/EIPs/issues/53. Accessed on 31 May 2017
42. Num TXs held in mempool. http://charts.bitcointicker.co/#mempooltrans
43. Blocke, J.: The Network Congestion Problem. 16 October 2016. https://keepingstock.net/network-congestion-is-problematic-c9d7829ed4ec#.d91t6vk9u
44. Hearn, M.: The Resolution of the Bitcoin Experiment. 14 January 2016. https://medium.com/@octskyward/the-resolution-of-the-bitcoin-experiment-dabb30201f7#.c5eb0hctq. Accessed on 31 May 2017
45. Timejacking & Bitcoin: 25 May 2011. http://culubas.blogspot.com/2011/05/timejacking-bitcoin_802.html
46. Ernst, D.: Problems. 23 May 2016. https://github.com/ethereum/wiki/wiki/Problems
47. EYAL, I.: The Miner's Dilemma. In: 2015 IEEE Symposium on Security and Privacy, San Jose, CA, pp. 89–103. (2015)
48. Karame, G. O., Androulaki, E., Roeschlin, M., Gervais, A., Capkun, S.: Misbehaviorin bitcoin: a study of double-spending and accountability. ACM Trans. Inf. Sys. Secur. 18(1), 21–32 (May 2015)

An Efficient Signature
Scheme for Anonymous Credentials

Chien-Nan Wu, Chun-I Fan, Jheng-Jia Huang, Yi-Fan Tseng
and Hiroaki Kikuchi

Abstract After Chaum introduced the concept of anonymous credential systems, a number of similar systems have been proposed; however, these systems use zero-knowledge protocols to authenticate users resulting in inefficient authentication in the possession of credential proving stage. In order to overcome this drawback of anonymous credential systems, we use partially blind signatures and chameleon hash functions to propose a signature scheme such that both the prover and the verifier achieve efficient authentication. In addition to giving a computational cost comparison table to show that the proposed signature scheme achieves more efficient possession proving than others, we provide concrete security proofs under the random oracle model to demonstrate that the proposed scheme satisfies the properties of anonymous credentials.

Keywords Anonymous credentials · Partially blind signatures · Chameleon hash functions · Information security

C.-N. Wu · C.-I. Fan (✉) · J.-J. Huang · Y.-F. Tseng
Department of Computer Science and Engineering, National Sun Yat-sen University,
Kaohsiung, Taiwan
e-mail: cifan@mail.cse.nsysu.edu.tw

C.-N. Wu
e-mail: eaze@ms46.hinet.net

J.-J. Huang
e-mail: jhengjia.huang@gmail.com

Y.-F. Tseng
e-mail: yftseng1989@gmail.com

H. Kikuchi
Department of Frontier Media Science, Meiji University, Surugadai, Japan
e-mail: kikn@meiji.ac.jp

© Springer International Publishing AG 2018
R. Lee (ed.), *Applied Computing & Information Technology*,
Studies in Computational Intelligence 727, DOI 10.1007/978-3-319-64051-8_4

1 Introduction

The concept of anonymous credential systems was introduced by Chaum [9] in 1985 in which it is a method of allowing the user to work effectively and anonymously with several organizations. In the anonymous credential system, a user can obtain a credential for a pseudonym from a service provider and then use it to access resources by demonstrating possession of the credential efficiently and anonymously without revealing any unnecessary information except for the fact that the credential is owned by the user. Using Chaum's underlying concept, many proposals [6–8, 10–12, 16] geared towards improving the model of the anonymous credential systems along with the zero-knowledge proofs have been presented in the literature. In the anonymous credential system, after negotiating with the signer (service provider) once to obtain the signature (credential), the user can use it unlimitedly to access the service provider's resource and thus, this structure is suitable for use in service-oriented architecture applications [2, 4].

The anonymous credential system has two properties: *unforgeability*: a credential is a signature and it cannot be forged; *unlinkability*: given two or more credentials, no one can identify whether these credentials are related to each other. That is, unlinkability ensures that users may make multiple uses of resources or services and these users and/or credentials are unable to infer whether the same user caused certain specific operations in the system.

1.1 Our Contributions and Paper Organization

1. All existing anonymous credential systems are inefficient in terms of possession of credential proving because they employ zero-knowledge proof protocols. In order to overcome this drawback, we propose a signature scheme for anonymous credential systems based on partially blind signatures and chameleon hash functions, which provides an expiration for the credential to limit the user's usage.
2. In addition to presenting a computational cost comparison table to show that the performance of the proposed scheme's possession of credential proving procedure is superior to that of other schemes, we provide concrete security proofs under the random oracle model to demonstrate that the proposed scheme satisfies the required properties for anonymous credential systems.

The remainder of this paper is organized as follows. Section 2 reviews preliminaries including the hard problems that are used in the proposed scheme and some related research. Section 3 provides the framework and the security models for a signature scheme with a provided expiration date. Then, Sect. 4 introduces the proposed scheme, Sect. 5 presents the proofs of security and a comparison on computation cost between our scheme and other similar ones. The final section presents the concluding remarks.

2 Preliminaries

2.1 Hard Problems

Let $\mathbb{G}$ be a cyclic group of prime order p and P be its generator. We review the hard problems used in this manuscript as follows.

1. **Discrete Logarithm (DL) Problem**:
   Given $(P, Q) \in \mathbb{G}$, find an integer $x \in \mathbb{Z}_p$ such that $Q = xP$.
2. **Extension of Modified Discrete Logarithm (E-MDL) Problem**:
   Given the instance $(P, aP, bP, caP) \in \mathbb{G}^3$, where $(a, b, c) \in \mathbb{Z}_p^3$, output a tuple $(v, \alpha_1, \alpha_2) \in \mathbb{Z}_p^3$ such that $vaP = \alpha_1 P - \alpha_2 bP - caP$.

Theorem 1 *The discrete logarithm (DL) problem is hard if and only if the extension of modified discrete logarithm $(E - MDL)$ problem is hard.*

Proof

- E-MDL $\propto$ DL
  If there exists an oracle which can solve the DL problem with non-negligible probability, then we can use the oracle to solve the E-MDL problem with non-negligible probability.

  – Take an E-MDL tuple (P, aP, bP, caP) as inputs, the DL oracle can output the corresponding values (a, b, ca) and further compute $v = \alpha_1 a - \alpha_2 ab - c$ mod p using arbitrary α_1 and α_2 as follows.

- DL $\propto$ E-MDL
  Assume that there exists an oracle which can solve the E-MDL problem with non-negligible probability. Thereafter, we demonstrate that the oracle can help us to solve the DL problem with non-negligible probability.

  – For DL tuple (P, aP), we input the parameters to the E-MDL oracle in the following steps.

    1. Choose $(b, c) \in \mathbb{Z}_p^2$ at random and compute bP and caP.
    2. Take (bP, caP) as inputs, the E-MDL oracle outputs a tuple (v, α_1, α_2).

  Thus, we retrieve $a = \frac{\alpha_1 - \alpha_2 b}{v + c}$ mod p.

2.2 Chameleon Hash Family

Based on the definition in [15], a chameleon hash family consists of two algorithms $(\mathscr{I}, \mathscr{H})$, where

1. $\mathcal{I}$ is a probabilistic polynomial-time key generation algorithm. On input a security parameter 1^k, it outputs a key pair (HK, TK) such that the sizes of HK and TK are polynomially related to k, in which HK is the hash key and TK is the trapdoor key.
2. $\mathcal{H}$ is a family of randomized hash functions. Every hash function in $\mathcal{H}$ is associated with a hash key HK, and its input format is a message from a space $\mathcal{M}$ and a random element from a finite space $\mathcal{R}$. Furthermore, the output of the hash function H_{HK} and the hash key TK are independent.

A chameleon hash family $(\mathcal{I}, \mathcal{H})$ has the following properties:

1. **Efficiency**: There exists an algorithm to compute $H_{HK}(m, r)$ in polynomial time by inputting a pair $(m, r) \in \mathcal{M} \times \mathcal{R}$ and a hash key HK.
2. **Collision Resistance**: On input HK, no probabilistic polynomial time algorithm $\mathcal{A}$ can output two pairs $(m_1, r_1), (m_2, r_2) \in \mathcal{M} \times \mathcal{R}$ with a non-negligible probability such that $m_1 \neq m_2$ and $H_{HK}(m_1, r_1) = H_{HK}(m_2, r_2)$ in which the probability is over HK, $(HK, TK) \leftarrow \mathcal{I}(1^k)$, and over the random coin tosses of algorithm $\mathcal{A}$.
3. **Trapdoor Collisions**: Given a key pair (HK, TK) and a pair $(m_1, r_1) \in \mathcal{M} \times \mathcal{R}$ with a message $m_2 \in \mathcal{M}$, there exists a probabilistic polynomial time algorithm can output a value $r_2 \in \mathcal{R}$ such that $H_{HK}(m_1, r_1) = H_{HK}(m_2, r_2)$ in which if r_1 is uniformly distributed in a finite space $\mathcal{R}$ then the distribution of r_2 is computationally indistinguishable from uniform in $\mathcal{R}$.

3 Frameworks

Definition 1 **An Efficient Signature Scheme with Expiration Date** $(\mathcal{ESSED})$. A signature scheme provided expiration date for anonymous credential includes two parties, a group of users and a signer, to participate in it and consists of the algorithms of **KeyGen, CHKeyGen, SignCH, SReq, SignCR, UnblindS, Verify,** and **SProving**. The details of those algorithms are described below.

1. $(PK, SK) \leftarrow$ **KeyGen**(k): Given a security parameter k, the probabilistic algorithm **KeyGen**(k) outputs a public-private key pair (PK, SK).
2. $(HK, TK) \leftarrow$ **CHKeyGen**(k): This algorithm **CHKeyGen**(k) outputs a hash-trapdoor key pair (HK, TK) by inputting the security parameter k.
3. $\bar{\sigma} \leftarrow$ **SignCH**(H, SK): Given a chameleon hash value H and the signing key SK, the **SignCH** algorithm outputs a signature $\bar{\sigma}$ on H.
4. $\alpha \leftarrow$ **SReq**(m, π, e, β): For a message m and a predefined information π (e.g. an expiration date), on inputting random secrets (e, β) chosen by the user, the **SReq** algorithm outputs a blinded message α, in which e and β can be regarded as a blinding factor and a randomized factor, respectively.

5. $\hat{r} \leftarrow$ **SignCR** (α, y, SK): Input the blinded message α, a random secret y selected by the signer, and the signer's secret key SK, the **SignCR** algorithm generates a blind signature $\hat{r}$.

6. $r \leftarrow$ **UnblindS** $(\hat{r}, e)$: Take a blind signature $\hat{r}$ and a blinding factor e as the inputs, the algorithm outputs the signature r. Therefore, the signature-message tuple is $\sigma = (r, m, \pi)$.

7. $\{Valid, \perp\} \leftarrow$ **Verify** (σ, PK, HK): Given a signature-message tuple σ, the deterministic algorithm outputs "*Valid*" if σ is a valid signature-message tuple or "$\perp$" otherwise.

8. $\{Valid, \perp\} \leftarrow$ **SProving** (σ', PK, HK): Given a new signature-message tuple σ' generated from σ, the deterministic algorithm **SProving** outputs "*Valid*" if σ' is a valid signature-message tuple or "$\perp$" otherwise.

3.1   Security Models

This section defines security models of unlinkability and unforgeability in the proposed signature scheme with a provided expiration date.

Definition 2 **Linkage Game**. Let U_0 and U_1 be two honest users and $\breve{S}$ be a signer of a signature scheme with a provided expiration date. $\breve{S}$ interacts with U_0 and U_1 by executing the following game.

Step 1. $\breve{S}$ performs the algorithms of **KeyGen**(k) and **CHKeyGen**(k), respectively, to obtain public-private key pair (PK, SK) and hash-trapdoor key pair (HK, TK), respectively. After that, $\breve{S}$ publishes system parameters for the signature scheme.

Step 2. $\breve{S}$ produces a negotiated information π and two distinct messages, m_0 and m_1.

Step 3. Randomly choose a bit $b \in \{0, 1\}$ and then input (m_b, π) and (m_{1-b}, π) into U_0's and U_1's private input tapes, respectively, where b is unknown to $\breve{S}$.

Step 4. $\breve{S}$ executes the signature scheme with U_0 and U_1, respectively.

Step 5. If U_0 and U_1 output two valid signature-message tuples σ_b and σ_{1-b} corresponding to m_b and m_{1-b} on their private input tapes, respectively, the two signature-message tuples are given to $\breve{S}$ in a random order; otherwise, $\perp$ is given to $\breve{S}$.

Step 6. $\breve{S}$ outputs $b' \in \{0, 1\}$ as the guess of b and $\breve{S}$ wins the game if $b' = b$.

Now we define $\breve{S}$'s advantage winning above game as $Adv_{\mathcal{CLSED}}^{Linkability}(\breve{S}) = \left| Pr[b' = b] - \frac{1}{2} \right|$ where $Pr[b' = b]$ denotes the probability of $b' = b$.

Definition 3 **Unlinkability**. A signature scheme with expiration date satisfies the property of unlinkability if $Adv_{\mathscr{CLSED}}^{Linkability}(\breve{S})$ in the Linkage Game is negligible.

Definition 4 **Forgery Game**. Let $\mathscr{A}$ be a probabilistic polynomial-time Turing machine and it interacts with a challenger $\mathscr{B}$ in the following game.

1. **Setup**(1^k):
   The challenger $\mathscr{B}$ selects a security parameter k and runs **KeyGen**(k) and **CHKeyGen**(k), respectively, to generate public-private key pair (PK, SK) and hash-trapdoor key pair (HK, TK), respectively. $\mathscr{B}$ also prepares system parameters and publishes those parameters to $\mathscr{A}$.
2. **Oracle Query**: $\mathscr{A}$ can make the queries to obtain some parameters and the details are described below.

   - $\mathcal{O}_{Request}$: $\mathscr{B}$ responds required parameters to $\mathscr{A}$ after receiving the query with a predefined information π from $\mathscr{A}$.
   - $\mathcal{O}_{Hash}$: This oracle is used to reply the hash queries from $\mathscr{A}$.
   - $\mathcal{O}_{Sign}$: $\mathscr{B}$ outputs a signature when $\mathscr{A}$ makes this query.

Finally, $\mathscr{A}$ produces a signature $\tilde{\sigma}$ on a message $\tilde{m}$ after querying those oracles and it wins the game if the following conditions hold:

1. The result of **Verify** $(\tilde{\sigma}, PK, HK)$ is valid.
2. $\mathscr{A}$ makes no signing query on the message $\tilde{m}$.

We define $\mathscr{A}$'s success advantage in the Forgery Game as $Adv_{\mathscr{CLSED}}^{Forgeability}(\mathscr{A})$.

Definition 5 **Unforgeability**. A signature scheme supplied expiration date meets the unforgeability property if $Adv_{\mathscr{CLSED}}^{Forgeability}(\mathscr{A})$ in the Forgery Game is negligible.

4   The Proposed Scheme

We introduce the proposed signature scheme with expiration date over the elliptic curve cryptosystem [13, 18] in this section. In order to raise the efficiency of the signing procedure, the proposed scheme's environment can be divided into on-line phase and off-line phase. In the proposed scheme, a group of users and the signer are involved and its details are presented as follows.

1. **Initialization Phase**. On inputting a security parameter k, the signer produces a prime p and a cyclic group $\mathbb{G}$ generated by P with order p. Let $(\mathscr{G}, \mathscr{S}ign, \mathscr{V}erify)$ be any provably secure signature scheme. After running the key generation algorithm $\mathscr{G}$ to acquire signing key SK and verifying key PK, the signer prepares two large primes $(x_1, x_2) \underset{\leftarrow}{R} \mathbb{Z}_p^2$ as the long-term trapdoor keys $TK = (x_1, x_2)$ of the chameleon hash family and then generates the corresponding hash keys $HK = (Y_1 = x_1 P, Y_2 = x_1 x_2^{-1} P)$. Furthermore, an employed

chameleon hash function is defined as $H_{HK} : \mathbb{Z}_p \times \mathbb{Z}_p \times \mathbb{Z}_p \times \mathbb{G} \times \mathbb{G} \to \mathbb{G}$, in which a secure one-way hash function used in H_{HK} is denoted as that $f : \mathbb{Z}_p \times \mathbb{Z}_p \times \mathbb{G} \times \mathbb{G} \to \mathbb{Z}_p$. Thereafter, the signer randomly chooses $\hat{k} \in \mathbb{Z}_p$ and sets the chameleon hash value $H = \hat{k}Y_1$, and then runs the signing algorithm $\mathscr{S}ign$ with the signing key SK to sign the message H, in which the signature is $\bar{\sigma} = \mathscr{S}ign_{SK}(H)$ and anyone can check its validity through the $\mathscr{V}erify$ algorithm by using the verifying key PK. Finally, the secret parameters and the public parameters are $SPar = (SK, TK, \hat{k})$ and $PPar = (p, P, \mathbb{G}, f, PK, HK, \bar{\sigma})$, respectively.

2. **Off-line Phase**. In this phase, both the user and the signer can prepare some parameters used in the on-line phase.

- **User Side**. After selecting a random number $\lambda \in \mathbb{Z}_p$ and determining two parameters (e, d) such that $ed \equiv 1 \pmod{p}$, the user computes λY_1 and then records the tuple $(e, d, \lambda, \lambda Y_1)$ into a list.
- **Signer Side**. The signer uses the random parameters $(z, y, t) \in \mathbb{Z}_p^3$ to compute the parameters $\theta = \hat{k} - zx_1^{-1} \bmod p$, zP, yY_1, and tY_2 and then stores these parameters as a tuple $(\theta, z, y, t, zP, yY_1, tY_2)$ into a list, where she/he can arrange multiple tuples at the same time.

3. **On-line Phase**

Step 1: Signature Requesting Phase. When a user asks for the service by sending a request with a predefined information π, the signer checks that whether π is of the predefined format. If yes, she/he picks a random tuple $(\theta, z, y, t, zP, yY_1, tY_2)$ and computes $\bar{\theta} = \theta - \pi x_1^{-1} \bmod p$. After that, the signer replies $(\bar{\theta}, zP, yY_1, tY_2)$ to the user. For the purpose of that the signer has no information about the signed message m, the user generates the following blind messages by means of the parameters $(e, d, \lambda, \lambda Y_1)$ and $(\bar{\theta}, zP, yY_1, tY_2)$.

$$\alpha_1 = (\bar{\theta} - \lambda)e \bmod p \tag{1}$$

$$\alpha_2 = f(m, \pi, yY_1, W)e + \lambda \bmod p \tag{2}$$

in which $W = zP + \lambda Y_1 + d\lambda yY_1 + dtY_2$. Thereafter, she/he submits (α_1, α_2) to the signer.

Step 2: Signing Phase. The signer uses her/his secret key x_2 and the secret parameters (y, t) to sign the blinded messages (α_1, α_2) as below.

$$\hat{r} = \alpha_1 x_2 - \alpha_2 yx_2 - t \bmod p \tag{3}$$

After that, she/he returns $\hat{r}$ with the user-trapdoor key yx_2 to the user through a secure channel.

Step 3: Signature Recovering Phase. After acquiring $\hat{r}$, the user takes the parameter d to extract the signature component r as:

$$r = \hat{r}d \mod p \tag{4}$$

Finally, the message-signature tuple is $\sigma = (m, \pi, r, yY_1, W)$ and its validity can be checked through the following verification equation.

$$f(m, \pi, yY_1, W)yY_1 + W + \pi P + rY_2 \stackrel{?}{=} \hat{k}Y_1 \tag{5}$$

Notably, σ will be not revealed to others.

4. **Signature Proving Phase**. In order to prove the knowledge of the signature $\sigma = (m, \pi, r, yY_1, W)$, the user, for a new message $\tilde{m}$, selects $(\beta, \delta) \in \mathbb{Z}_p^2$ at random and then utilizes the trapdoor key yx_2 to compute the following parameters.

$$\tilde{W} = W + \delta Y_2 \tag{6}$$

$$\tilde{r} = r + f'yx_2 - \tilde{f}\beta yx_2 - \delta \mod p \tag{7}$$

in which $f' = f(m, \pi, yY_1, W)$ and $\tilde{f} = f(\tilde{m}, \pi, \beta yY_1, \tilde{W})$. Thus, the new message-signature tuple is $\sigma = (\tilde{m}, \pi, \tilde{r}, \beta yY_1, \tilde{W})$ and its verification equation is presented below.

$$f(\tilde{m}, \pi, \beta yY_1, \tilde{W})\beta yY_1 + \tilde{W} + \pi P + \tilde{r}Y_2 \stackrel{?}{=} \hat{k}Y_1 \tag{8}$$

5 Security Proofs and Performance Analysis

5.1 Security Proofs

We demonstrate that our proposed signature scheme provides unlinkability and unforgeability properties in this part and the details are presented as follows.

- **Property: Unlinkability**

Theorem 2 *The proposed signature scheme satisfies the unlinkability property.*

Proof If $\breve{S}$ is given $\bot$ in Step 5 of the Linkage Game, it means that $\breve{S}$ cannot obtain additional information about message-signature tuples σ_b and σ_{1-b} under the same predefined information π, and therefore, she/he can only determine a random guess of b with the probability $\frac{1}{2}$.

Otherwise, let $(\pi, \bar{\theta}_i, z_i P, y_i Y_1, t_i Y_2, \alpha_{1,i}, \alpha_{2,i}, \hat{r}_i, y_i x_2)$, $i \in \{0,1\}$, be the view of parameters exchanged during the signature procedure of our proposed signature scheme. Assume that $\breve{S}$ can get two signature-message tuples $(\pi, \tilde{m}_i, \tilde{r}_i, \beta_i y_i Y_1, \tilde{W}_i)$, $i \in \{0,1\}$, after running the signature proving stage of the proposed signature protocol. For a given tuple $(\pi, \tilde{m}, \tilde{r}, \beta y Y_1, \tilde{W}) \in \{(\pi, \tilde{m}_0, \tilde{r}_0, \beta_0 y_0 Y_1, \tilde{W}_0)(\pi, \tilde{m}_1, \tilde{r}_1,, \beta_1 y_1 Y_1, \tilde{W}_1)\}$ and a view $(\pi, \bar{\theta}_i, z_i P, y_i Y_1, t_i Y_2, \alpha_{1,i}, \alpha_{2,i}, \hat{r}_i)$, $i \in \{0,1\}$, there always exists a corresponding tuple $(\lambda'_i, e'_i/d'_i, \beta_i, \delta_i)$ such that the formulas (1)–(8) hold.

For the given tuples $(\pi, \tilde{m}, \tilde{r}, \beta y Y_1, \tilde{W})$ and $(\pi, \bar{\theta}_i, z_i P, y_i Y_1, t_i Y_2, \alpha_{1,i}, \alpha_{2,i}, \hat{r}_i, y_i x_2)$ through the formulas (1)–(3) and (6), there always exists $(\lambda'_i, e'_i/d'_i, \beta'_i, \delta'_i)$ such that the following equations are held.

$$\tilde{W} = z_i P + \lambda'_i Y_1 + d'_i \lambda'_i y_i Y_1 + d'_i t_i Y_2 + \delta'_i Y_2$$
$$\alpha_{1,i} = (\bar{\theta} - \lambda_{i'}) e'_i \bmod p$$
$$\alpha_{2,i} = f(m_i, \pi, y_i Y_1, W_i) e_{i'} + \lambda'_i \bmod p$$

From the formula (7), we can get

$$
\begin{aligned}
\tilde{r} \quad &\equiv r_i + f(m_i, \pi, y_i Y_1, W_i) y_i x_2 - f(\tilde{m}_i, \pi, \beta_i y_i Y_1, \tilde{W}_i) \beta'_i y_i x_2 - \delta'_i \\
&\equiv \hat{r}_i d'_i + f(m_i, \pi, y_i Y_1, W_i) y_i x_2 - f(\tilde{m}_i, \pi, \beta_i y_i Y_1, \tilde{W}_i) \beta'_i y_i x_2 - \delta'_i \\
&\equiv (\alpha_{1,i} x_2 - \alpha_{2,i} y_i x_2 - t_i) d_{i'} + f(m_i, \pi, y_i Y_1, W_i) y_i x_2 \\
&\quad - f(\tilde{m}_i, \pi, \beta_i y_i Y_1, \tilde{W}_i) \beta'_i y_i x_2 - \delta'_i \\
&\equiv ((\bar{\theta}_i - \lambda'_i) e'_i x_2 - (f(m_i, \pi, y_i Y_1, W_i) e'_i + \lambda'_i) y_i x_2 - t_i) d'_i \\
&\quad + f(m_i, \pi, y_i Y_1, W_i) y_i x_2 - f(\tilde{m}_i, \pi, \beta_i y_i Y_1, \tilde{W}_i) \beta'_i y_i x_2 - \delta'_i \\
&\equiv ((\theta - \pi x_1^{-1} - \lambda'_i) e'_i x_2 - (f(m_i, \pi, y_i Y_1, W_i) e'_i + \lambda'_i) y_i x_2 - t_i) d'_i \\
&\quad + f(m_i, \pi, y_i Y_1, W_i) y_i x_2 - f(\tilde{m}_i, \pi, \beta_i y_i Y_1, \tilde{W}_i) \beta'_i y_i x_2 - \delta'_i \\
&\equiv ((\hat{k} - z_i x_1^{-1} - \pi x_1^{-1} - \lambda'_i) e'_i x_2 - (f(m_i, \pi, y_i Y_1, W_i) e'_i + \lambda'_i) y_i x_2 \\
&\quad - t_i) d'_i + f(m_i, \pi, y_i Y_1, W_i) y_i x_2 - f(\tilde{m}_i, \pi, \beta_i y_i Y_1, \tilde{W}_i) \beta'_i y_i x_2 - \delta'_i \\
&\equiv (\hat{k} - z_i x_1^{-1} - \pi x_1^{-1} - \lambda'_i) x_2 - f(\tilde{m}_i, \pi, \beta_i y_i Y_1, \tilde{W}_i) \beta'_i y_i x_2 - d'_i \lambda'_i y_i x_2 \\
&\quad - d'_i t_i - \delta'_i \pmod{p}
\end{aligned}
$$

and thus the result implies that the formula (8) always holds.

Therefore, there always exists a tuple $(\lambda'_i, e'_i/d'_i, \beta'_i, \delta'_i)$ such that the formulas (1)–(8) are held by giving any signature-message tuple $(\tilde{m}, \pi, \tilde{r}, \beta y Y_1, \tilde{W}) \in \{(\tilde{m}_0, \pi, \tilde{r}_0, \beta_0 y Y_1, \tilde{W}_0), (\tilde{m}_1, \pi, \tilde{r}_1, \beta_1 y Y_1, \tilde{W}_1)\}$ and any view $(\pi, \bar{\theta}_i, z_i P, y_i Y_1, t_i Y_2, \alpha_{1,i}, \alpha_{2,i}, \hat{r}_i, y_i x_2), i \in \{0,1\}$. As the result, the signer $\breve{S}$ would only succeed in determining b with probability $\frac{1}{2}$ in Step 6 of the game. According to the result, we have $Pr[b' = b] = \frac{1}{2}$ and $Adv_{\mathcal{ELSED}}^{Linkability}(\breve{S}) = 0$. Thus, the proposed signature scheme with an expiration date satisfies the unlinkability property. $\qquad\square$

- **Security: Unforgeability**

To prove Chaum's blind signature scheme against signature forgery, Bellare et al. [3] introduced a security assumption, *chosen-target RSA-inversion assumption*, and the details are described below.

Definition 6 **Chosen-Target Inversion Problem, alternative formulation (RSA-ACTI).** Let $k \in \mathbb{N}$ be a security parameter. Let $\mathscr{A}$ be an adversary with access to the RSA-inversion oracle $(\cdot)^d \bmod N$ and the challenge oracle $\mathcal{O}_N$. Consider the following experiment:

Experiment EXP $_{\mathscr{A}}^{rsa-acti}(k)$

$(N, e, d) \overset{R}{\leftarrow} \text{KeyGen } (k)$

$(\hat{\pi}, x_1, \ldots, x_m) \leftarrow \mathscr{A}^{(\cdot)^d \bmod N, \mathcal{O}_N}(N, e, k)$, where m is an integer
Let n be the number of queries to $\mathcal{O}_N$
Let $y_1, \ldots, y_n$ be the challenges returned by $\mathcal{O}_N$
If the following are all true then return 1 else return 0

- $\hat{\pi} : \{1, \ldots, m\} \to \{1, \ldots, n\}$ is injective
- $\forall i \in \{1, \ldots, m\} : x_i^e \equiv y_{\hat{\pi}(i)} \pmod N$
- $\mathscr{A}$ made strictly fewer than m queries to $(\cdot)^d \bmod N$

The advantage of $\mathscr{A}$ is $Adv_{\mathscr{A}}^{rsa-acti}(k) = Pr[EXP_{\mathscr{A}}^{rsa-acti}(k) = 1]$. The RSA-ACTI problem is said to be hard if the function $Adv_{\mathscr{A}}^{rsa-acti}(\cdot)$ is negligible for any adversary $\mathscr{A}$ whose time-complexity is polynomial in the security parameter k.

Based on the Definition 6, we propose a modified version of the RSA-ACTI in $\mathbb{G}$, where $\mathbb{G}$ is a cyclic additive group generated by P. It is explained as follows.

Definition 7 **Extension of Modified Chosen-Target Inversion Assumption (E-MCTI).** Let $\mathbb{G}$ be a group with prime order p generated by P. For the instance tuple $(P, a^{-1}P)$, two oracles are described below.

- Target oracle $\mathcal{O}_{\mathscr{T}}$:
  1. Select $(b, c) \overset{R}{\leftarrow} \mathbb{Z}_p$
  2. Return $(bP, ca^{-1}P)$ as the outputs

- Help oracle $\mathcal{O}_{\mathscr{H}}$:
  1. Take a tuple $(bP, ca^{-1}P)$ and values $(\alpha_1, \alpha_2) \in \mathbb{Z}_p$ as inputs, where the tuple $(bP, ca^{-1}P)$ is from $\mathcal{O}_{\mathscr{T}}$
  2. Output v such that $v = \alpha_1 a - \alpha_2 ab - c \bmod p$
  3. One restriction is that each of the same bP and $ca^{-1}P$ cannot be taken as the inputs twice with distinct tuples (α_1, α_2) and (α_1', α_2').

If there exists an adversary $\mathscr{A}$ with access to the target oracle $\mathcal{O}_{\mathscr{T}}$ and the help oracle $\mathcal{O}_{\mathscr{H}}$ that can output a set $V = \{(b_1P, c_1a^{-1}P, v_1, \alpha_{1,1}, \alpha_{2,1}) \cdots (b_\ell P, c_\ell a^{-1}P, v_\ell, \alpha_{1,\ell}, \alpha_{2,\ell})\}, q_{\mathscr{H}} < \ell \leq q_{\mathscr{T}}$, such that $v_i a^{-1}P = \alpha_{1,i}P - \alpha_{2,i}b_iP - c_ia^{-1}P, 1 \leq i \leq \ell$, after making $q_{\mathscr{F}}$ $\mathscr{O}_{\mathscr{F}}$ queries to obtain the following $q_{\mathscr{F}}$ tuples $\{(b_1P, c_1a^{-1}P), \ldots, (b_{q_{\mathscr{F}}}P, c_{q_{\mathscr{F}}}a^{-1}P)\} \in \mathbb{G}^{q_{\mathscr{F}}}$ and $q_{\mathscr{H}}$ $\mathscr{O}_{\mathscr{H}}$ queries $(q_{\mathscr{H}} < q_{\mathscr{F}})$, where the set V may be formed by the following formats: one is that all tuples $(b_iP, c_ia^{-1}P, v_i, \alpha_{1,i}, \alpha_{2,i})$, $1 \leq i \leq \ell$, are distinct; the other is that the $(b_\ell P, c_\ell a^{-1}P)$ is one of the tuple $(b_jP, c_ja^{-1}P)$, $1 \leq j \leq \ell - 1$ and $(v_\ell, \alpha_{2,\ell})$ is different from the tuples $\{(v_1, \alpha_{2,1}), \ldots, (v_{\ell-1}, \alpha_{2,\ell-1})\}$. Therefore, we can say $\mathscr{A}$ wins the game.

Definition 8 **One-More Forgery** [19]. For a signature scheme with any integer $\ell = polynomial(k)$, where k is a security parameter. An $(\ell, \ell+1)$-forgery means that a probabilistic polynomial-time Turing machine can output $\ell+1$ valid signatures with non-negligible probability after ℓ interactions with the signer. The one-more forgery is an $(\ell, \ell+1)$-forgery for some integer ℓ.

Lemma 1 *Forking Lemma [20]. Let (s, m, c) be a valid signature-message triple of a signature scheme and h be the hashed value of (m, c) where m is a plaintext message, c is a string, and s is called the signature part of the triple. Let A be a probabilistic polynomial-time Turning machine. Input the public data of the signature scheme, if A can find, with non-negligible probability, a valid signature-message triple (s, m, c) with h, then, with non-negligible probability, a replay of this machine, with the same random tape and a different value returned by the random oracle, outputs two valid signature-message triples of (s, m, c) with h and (s', m, c) with h' such that $h \neq h'$.*

Lemma 2 *Splitting Lemma [20]. Let $A \subset X \times Y$ such that $Pr[(x, y) \in A] \geq \varepsilon$. For any $\alpha < \varepsilon$, define $B = \{(x, y) \in X \times Y | Pr_{y' \in Y}[(x, y') \in A] \geq \varepsilon - \alpha\}$ and then the following conditions hold:*

1. $Pr[B] \geq \alpha$
2. $\forall (x, y) \in B, Pr_{y' \in Y}[(x, y') \in A] \geq \varepsilon - \alpha\}$
3. $Pr[B|A] \geq \alpha/\varepsilon$

Theorem 3 *The proposed $\mathscr{CLSED}$ is secure against one-more forgery under the extension of modified chosen-target inversion assumption.*

Proof Assume that $\mathscr{A}$ is an adversary succeeding in one-more forgery on the proposed $\mathscr{CLSED}$, and then we can construct a solver $\mathscr{B}$ to help us against the E-MCTI assumption by using $\mathscr{A}$'s capability. The scenario of simulation is presented in the following.

1. **Initiation.** For the challenge tuple $(P, a^{-1}P) \in \mathbb{G}^2$, the simulator randomly chooses two elements $(\hat{k}, x_1) \in \mathbb{Z}_p^2$ to set $HK = (Y_1 = x_1P, Y_2 = x_1a^{-1}P)$ as the public hash key and $H = \hat{k}Y_1$ as the chameleon hash value. After that, $\mathscr{B}$ simulates a one-way hash function such that $f : \mathbb{Z}_p \times \mathbb{Z}_p \times \mathbb{G} \times \mathbb{G} \rightarrow \mathbb{Z}_p$ and employs a provably secure signature scheme $(\mathscr{G}, \mathscr{Sign}, \mathscr{Verify})$, in which signing/verifying key pair of the signature scheme is (SK, PK). Therefore, the system public parameter is $(p, P, \mathbb{G}, f, PK, HK, \bar{\sigma})$, in which $\bar{\sigma}$ is the signature of

the chameleon hash value H signed by SK. Before interacting with $\mathscr{A}$ (in the off-line phase), $\mathscr{B}$ chooses $z \in \mathbb{Z}_p$ at random, and computes the zP and $\theta = \hat{k} - zx_1^{-1} \bmod p$. Here, $\mathscr{B}$ can arrange several tuples of (θ, zP) in this phase and stores them in a list. Besides, in order to respond to some queries from $\mathscr{A}$, $\mathscr{B}$ is permitted to access the target oracle $\mathscr{O}_{\mathscr{T}}$ and the help oracle $\mathscr{O}_{\mathscr{H}}$ offered in the E-MCTI assumption.

2. **Oracle Query**. $\mathscr{A}$ can make queries on a parameter request oracle $\mathscr{O}_{Request}$, two hash oracles $\mathscr{O}_{Hash_1}$ and $\mathscr{O}_{Hash_2}$, and a signature oracle $\mathscr{O}_{Sign}$ which the details of simulation are described in the following.

- $\mathscr{O}_{Request}$ query: When $\mathscr{A}$ makes this query with a predefined information π, $\mathscr{B}$ sends a request to the target oracle $\mathscr{O}_{\mathscr{T}}$ to acquire parameter tuple $(bP, ca^{-1}P)$. After receiving $(bP, ca^{-1}P)$, $\mathscr{B}$ computes the parameters $x_1bP, x_1ca^{-1}P$ and $\bar{\theta} = \theta - \pi x_1^{-1} \bmod p$ by using the secret key x_1 with a random tuple (θ, zP). Next, she/he responds the tuple $(\bar{\theta}, zP, x_1bP, x_1ca^{-1}P)$ to $\mathscr{A}$ and records it in a list $L_{Request}$.
- $\mathscr{O}_{Hash}$ query: $\mathscr{B}$ maintains a list L_{Hash} of tuple (m, π, f') and it is initially empty. When $\mathscr{A}$ makes a query on (m, π, x_1bP, W), $\mathscr{B}$ responds the corresponding value f' to $\mathscr{A}$ if (m, π, x_1bP, W) is in L_{Hash}; otherwise, $\mathscr{B}$ chooses $f'R \leftarrow \mathbb{Z}_p$ and returns it to $\mathscr{A}$, where $\mathscr{B}$ also stores (m, π, x_1bP, W, f') into the list L_{Hash}.
- $\mathscr{O}_{Sign}$ query: When $\mathscr{A}$ sends the tuple $(\alpha_1, \alpha_2, x_1bP, x_1ca^{-1}P)$ as inputs for acquiring the signature, $\mathscr{B}$ checks whether $(x_1bP, x_1ca^{-1}P)$ is in the list $L_{Request}$. If it is true, $\mathscr{B}$ queries the help oracle $\mathscr{O}_{\mathscr{H}}$ by inputting $(\alpha_1, \alpha_2, bP, ca^{-1}P)$ to obtain the value $v = \alpha_1 a - \alpha_2 ab - c \bmod p$ and then returns the result to $\mathscr{A}$; otherwise, $\mathscr{B}$ aborts it.

3. **Forgery and Problem Solving**. After making $q_{Request}$, q_{Hash}, and q_{Sign} queries on oracles of $\mathscr{O}_{Request}$, $\mathscr{O}_{Hash}$, and $\mathscr{O}_{Sign}$, respectively, if $\mathscr{A}$ can output ℓ valid signature-message tuples $(m_i, \pi_i, r_i, b_i x_1 P, W_i)$, $1 \leq i \leq \ell$, of the proposed signature scheme for $q_{Sign} < \ell \leq q_{Request}$, then $\mathscr{B}$ can fork the forged signature once obtaining two valid signatures $(\bar{m}, \bar{\pi}, \bar{r}, \bar{b}x_1P, \bar{W})$ and $(\bar{m}, \bar{\pi}, \tilde{r}, \bar{b}x_1P, \bar{W})$ with respect to the hash values $\bar{f}'$ and $\tilde{f}'$ such that $\bar{f}' \neq \tilde{f}'$ with the probability at least $\frac{\varepsilon}{2}$ by Lemma 1:

$$\begin{cases} \bar{f}'\bar{b}x_1P + \bar{W} + \bar{\pi}P + \bar{r}Y_2 = \hat{k}Y_1 \\ \tilde{f}'\bar{b}x_1P + \bar{W} + \bar{\pi}P + \tilde{r}Y_2 = \hat{k}Y_1 \end{cases}$$

Then, we compute $a\bar{b} = \frac{\tilde{r} - \bar{r}}{\bar{f}' - \tilde{f}'}$ with probability at least $\frac{\varepsilon}{2q_{Hash}} - \frac{1}{p^{\ell} \cdot q_{Sign}}$ in which we observe that $a\bar{b}$ is the user-trapdoor key; hence, we can find the corresponding tuple $(\bar{b}P, \bar{c}a^{-1}P, \bar{v}, \bar{\alpha}_{1,j}, \bar{\alpha}_{2,j})$ of $\bar{b}x_1P$ and then produce ℓ-th tuple $(\bar{b}P, \bar{c}a^{-1}P, \bar{v}', \bar{\alpha}_{1,j}, \bar{\alpha}'_{2,j})$ for a constant $\bar{\alpha}'_{2,j}$, where $\bar{v}' = \bar{v} + \bar{\alpha}_{2,j}a\bar{b} - \bar{\alpha}'_{2,j}a\bar{b}$. Finally, we output

$$\left\{\begin{array}{c} (b_1P, c_1a^{-1}P, v_1, \alpha_{1,1}, \alpha_{2,1}) \\ \vdots \\ (b_{\ell-1}P, c_{\ell-1}a^{-1}P, v_{\ell-1}, \alpha_{1,\ell-1}, \alpha_{2,\ell-1}) \\ (\bar{b}P, \bar{c}a^{-1}P, \bar{v}', \bar{\alpha}_{1,j}, \bar{\alpha}'_{2,j}) \end{array}\right\}$$

as ℓ valid instances of the E-MCTI assumption such that each tuple satisfies the equation: $v_i a^{-1}P = \alpha_{1,i}P - \alpha_{2,i}bP - c_i a^{-1}P, 1 \leq i \leq \ell$, in which $q_{\mathscr{H}} = q_{Sign} < \ell \leq q_{Request} = q_{\mathscr{T}}$ and $q_{\mathscr{H}}$ and $q_{\mathscr{T}}$ are the numbers of queries to the help oracle $\mathcal{O}_{\mathscr{H}}$ and the target oracle $\mathcal{O}_{\mathscr{T}}$, respectively. Obviously, it contradicts the E-MCTI assumption.

5.2 Performance Analysis

We sum up the comparison of computation costs between ours and some anonymous credential protocols [1, 5, 7, 8, 22] in Table 1 to show that our proposed signature scheme in the knowledge proving stage is extremely efficient. Note that we have deducted some of the pre-computations associated with these comparative schemes in order to compare fairly and let the attribute number in [5] be 1 in order to refer to the attribute of expiration date.

Table 1 Comparison of computation costs

	On-line signing phase	Signature proving phase
[1]	$\geq t_e$ $\approx 240t_m(0.619\,\text{ms})$	$\geq 2T_p$ $\approx 2400t_m(6.192\,\text{ms})$
[5]	$\approx 2Tm + 4t_m + t_h$ $\approx 63t_m(0.163\,\text{ms})$	$\approx 6T_m + 2t_m$ $\approx 166t_m(0.429ms)$
[7]	$\geq 7t_e$ $\approx 1680t_m(4.334\,\text{ms})$	$\geq 11t_e$ $\approx 2640t_m(6.811\,\text{ms})$
[8]	$\geq 2T_p$ $\approx 2400t_m(6.192\,\text{ms})$	$\geq 14T_p$ $\approx 16800t_m(43.344\,\text{ms})$
[22]	$\geq 2T_p$ $\approx 2400t_m(6.192\,\text{ms})$	$\geq 3T_p$ $\approx 3600t_m(9.288\,\text{ms})$
Ours	$\approx 2T_m + 6t_m + t_h$ $\approx 65t_m(0.168\,\text{ms})$	$\approx 4T_m + 3t_m + 3t_h$ $\approx 122t_m(0.315\,\text{ms})$

According to [21], timing of pairing operation is about 3.10 ms
According to [14, 17, 23]: $T_p \approx 5t_e$, $t_e \approx 240t_m$, $T_m \approx 29t_m$, $t_h \approx t_m$
Notations
T_p: The time cost of a pairing operation
T_m: The time cost of a scalar multiplication in G_1
t_e: The time cost of a modular exponentiation in Z_p
t_m: The time cost of a modular multiplication in Z_p
t_h: The time cost of a hash operation

6 Conclusions

On the basis of partially blind signatures and double-trapdoor chameleon hash functions, we presented a signature scheme with an expiration date provided under the elliptic curve primitives for anonymous credential systems. This improves the efficiency in the possession of credential proving stage in which the credential is restricted to a valid period. Further, we gave a computational cost comparison table to show that the proposed signature scheme is significantly more efficient than other methods in terms of possession of credential proving. Moreover, we provided concrete security proofs under the random oracle model to demonstrate that the proposed signature scheme satisfies the essential properties of anonymous credential systems.

Acknowledgements This work was partially supported by the Ministry of Science and Technology of the Taiwan under grants MOST 105-2923-E-110-001-MY3 and MOST 105-2221-E-110-053-MY2.

References

1. Akagi, N., Manabe, Y., Okamoto, T.: An efficient anonymous credential system. In: Proceedings of 12th International Conference on Financial Cryptography and Data Security—FC'08, LNCS 5143, pages 272–286. Springer, Berlin (2008)
2. Alodib, M.: Towards a monitoring framework for the automatic integration of the access control policies for web services. Int. J. Networked Distrib. Comput. **3**, 137–149 (2015)
3. Bellare, M., Namprempre, C., Pointcheval, D., Semanko, M.: The one-more-rsa-inversion problems and the security of Chaum's blind signature scheme. J. Cryptology **16**, 185–215 (2003)
4. Bouchiha, D., Malki, M., Djaa, D., Alghamdi, A., Alnafjan, K.: Towards a monitoring framework for the automatic integration of the access control policies for web services. Int. J. Networked Distrib. Comput. **2**, 35–44 (2014)
5. Brands, S.A.: Rethinking Public Key Infrastructures and Digital Certificates: Building in Privacy (2000)
6. Camenisch, J., Lysyanskaya, A.: Efficient non-transferable anonymous multi-show credential system with optional anonymity revocation. In Proceedings of Advances in Cryptology—EUROCRYPT'01, LNCS 2045, pp. 93–118. Springer, Berlin (2001)
7. Camenisch, J., Lysyanskaya, A.: A signature scheme with efficient protocols. In Proceedings of 3rd International Conference on Security in Communication Networks—SCN'02, LNCS 2576, pp. 268–289. Springer, Berlin (2003)
8. Camenisch, J., Lysyanskaya, A.: Signature schemes and anonymous credentials from bilinear maps. In: Proceedings of Advances in Cryptology—CRYPTO'04, LNCS 3152, pp. 56–72. Springer, Berlin (2004)
9. Chaum, D.: Security without identification: transaction systems to make big brother obsolete. Commun. ACM **28**(10), 1030–1044 (1985)
10. Chaum, D., Evertse, J.H.: A secure and privacy-protecting protocol for transmitting personal information between organizations. In: Proceedings on Advances in Cryptology—CRYPTO'86, LNCS 263, pp. 118–167. Springer, Berlin (1987)
11. Chen, L.: Access with pseudonyms. In: Proceedings of International Conference on Cryptography: Policy and Algorithms, LNCS 1029, pp. 232–243. Springer, Berlin (1996)

12. Damgard, I.B.: Payment systems and credential mechanisms with provable security against abuse by individuals. In: Proceedings of Advances in Cryptology—CRYPTO'88, LNCS 403, pp. 328–335. Springer, Berlin (1990)
13. Koblitz, N.: Elliptic curve cryptosystems. Math. Comput. **48**(177), 203–209 (1987)
14. Koblitz, N., Menezes, A., Vanstone, S.: The state of elliptic curve cryptography. Des. Codes Crypt. **19**(2–3), 173–193 (2000)
15. Krawczyk, H., Rabin, T.: Chameleon signatures. In: Proceedings of Network and Distributed System Security Symposium—NDSS'00, pp. 143–154 (2000)
16. Lysyanskaya, A., Rivest R.L., Sahai, A., Wolf, S.: Pseudonym systems. In Proceedings of 6th Annual International Workshop on Selected Areas in Cryptography—SAC'99, LNCS 1758, pp. 184–199. Springer, Berlin (2000)
17. Menezes, A.J., Vanstone, S.A., Van Oorschot, P.C.: Handbook of Applied Cryptography, 1st edn. CRC Press Inc, Boca Raton, FL, USA (1997)
18. Miller, VS.: Use of elliptic curves in cryptography. In Proceedings of Advances in Cryptology—CRYPTO'85, LNCS 218, pp. 417–426. Springer, Berlin (1986)
19. Pointcheval, D., Stern, J.: Provably secure blind signature schemes. In: Proceedings of Advances in Cryptology—ASIACRYPT'96, LNCS 1163, pp. 252–265. Springer, Berlin (1996)
20. Pointcheval, D., Stern, J.: Security proofs for signature schemes. In: Proceedings of Advances in Cryptology—EUROCRYPT'96, LNCS 1070, pp. 387–398. Springer, Berlin (1996)
21. Scott, M.: Implementing cryptographic pairings. In: Proceedings of the First international conference on Pairing-Based Cryptography, Pairing'07, pp. 177–196. Springer, Berlin (2007)
22. Tsang, P.P., Au, M.H., Kapadia, A., Smith, S.W.: Blacklistable anonymous credentials: blocking misbehaving users without ttps. In: Proceedings of the 14th ACM Conference on Computer and Communications Security—CCS'07, pp. 72–81. ACM (2007)
23. Zhang, Y., Liu, W., Lou, W., Fang, Y.: Securing mobile ad hoc networks with certificateless public keys. IEEE Trans. Dependable Secure Comput **3**(4), 386–399 (2006)

Improve Example-Based Machine Translation Quality for Low-Resource Language Using Ontology

Khan Md Anwarus Salam, Setsuo Yamada and Nishio Tetsuro

Abstract In this research we propose to use ontology to improve the performance of an EBMT system for low-resource language pair. The EBMT architecture use (CSTs) and unknown word translation mechanism. CSTs consist of a chunk in source-language, a string in target-language, and word alignment information. For unknown word translation, we used WordNet hypernym tree and English-Bengali dictionary. CSTs improved the wide-coverage by 57 points and quality by 48.81 points in human evaluation. Currently 64.29% of the test-set translations by the system were acceptable. The combined solutions of CSTs and unknown words generated 67.85% acceptable translations from the test-set. Unknown words mechanism improved translation quality by 3.56 points in human evaluation.

Keywords Knowledge engineering · WordNet · Example-Based machine translation

1 Introduction

Example-Based Machine Translation (EBMT) for low-resource language pair, like English-Bengali, has low-coverage issues, due to the lack of parallel corpus. It also has high probability to handle unknown words.

K.M.A. Salam (✉)
IBM Research, Tokyo, Japan
e-mail: khan@jp.ibm.com

S. Yamada
NTT Corporation, Tokyo, Japan
e-mail: yamada.setsuo@lab.ntt.co.jp

N. Tetsuro
Graduate School of Informatics and Engineering, The University of Electro-Communications, Tokyo, Japan
e-mail: nishino@uec.ac.jp

© Springer International Publishing AG 2018
R. Lee (ed.), *Applied Computing & Information Technology*,
Studies in Computational Intelligence 727, DOI 10.1007/978-3-319-64051-8_5

In this research we propose to use ontology to improve the performance of an EBMT system for low-resource language pair. The EBMT architecture use chunk-string templates (CSTs) and unknown word translation mechanism. Using WordNet [1] CSTs help to achieve wide-coverage and better quality in EBMT for low-resource language pair like English to Bengali language. For unknown word translation, we used related information such as synsets and hypernyms from WordNet.

CSTs consist of a chunk in source-language, a string in target-language, and word alignment information. CSTs are prepared automatically from word aligned parallel corpus. First the source-language chunks are auto generated by using OpenNLP chunker. Then initial CSTs are generated for each source-language chunks and each CSTs alignment for all target sentences are generated using the parallel corpus. After that the system generates combinations of CSTs using the word alignment information. Finally, we generalize CSTs using Word-Net to achieve wide-coverage.

For unknown word translation, we used WordNet hypernym tree and English-Bengali dictionary. At first the system finds the set of hyper-nyms words and degree of distance from the English WordNet. Then the system tries to find the translation of hypernym words from the dictionary according to the degree of distance order. When no dictionary entry found from the hypernym tree, it transliterates the word.

We built an English-to-Bengali EBMT system using CSTs and un-known word translation mechanism. CSTs improved the wide-coverage by 57 points and quality by 48.81 points in human evaluation. Currently 64.29% of the test-set translations by the system were acceptable. Unknown words mechanism improved translation quality by 3.56 points in human evaluation. The combined solutions of CSTs and un-known words generated 67.85% acceptable translations from the test-set.

The rest of this article is organized as follows. In Sect. 2 we explain the background research of this paper. Section 3 gives a brief overview of our proposed EBMT Architecture. Then in Sect. 4 we explain the CSTs generation process and usage method in translation in details. For reporting the evaluation result we describe our findings in Sect. 5. Then in Sect. 6 we discuss our findings. Then finally we conclude our paper in Sect. 7.

2   Background

Bengali is the native language of around 230 million people world-wide, mostly from Bangladesh. According to "Human Development Report 2009" of United Nations Development Program, the literacy rate of Bangladesh is 53.5%. So we can assume that around half of Bengali speaking people are monolingual. Since significant amount of the web contents are in English, it is important to have English to Bengali Machine Translation (MT) system. But English and Bengali form a distant language pair, which makes the development of MT system very challenging. Bengali is considered as low-resource language, due to the lack of language resources like electronic texts and parallel corpus. As a result, most of the commercial MT systems do not support Bengali language translation.

In present, there are several ways of (MT) such as Rule-Based Machine Translation (RBMT), Statistical Machine Translation (SMT) and Example-Based Machine Translation (EBMT) which includes chunk-based and template-based approaches.

RBMT require human made rules, which are very costly in terms of time and money, but still unable to translate general-domain texts. There are several attempts in building English-Bengali MT system. The first available free MT system from Bangladesh was Akkhor Bangla Software. The second available online MT system was apertium based Anubadok. These systems used Rule-Based approach and did not con-sider about improving translation coverage by handling unknown words, in low-resource scenario. Dasgupta et al. [2] proposed to use syntactic transfer. They converted CNF trees to normal parse trees and using a bilingual dictionary, generated output translation. This research did not consider translating unknown words.

SMT works well for close language pairs like English and French. It requires huge parallel corpus, but currently huge English-Bengali parallel corpus is not available. The most widely used SMT system for Bengali language is Google translate. Google Translate is a free translation service that provides instant translations between dozens of different languages. It can translate words, sentences and web pages between any combination of our supported languages. However, the quality of the translation produced heavily suffer from unknown words problem. English to Bengali phrase-based (SMT) was reported by Islam et al. [3]. This system achieved low BLEU score due to small parallel corpus for English-Bengali.

EBMT is better choice for low-resource language, because we can easily add linguistic information into it. Comparing with SMT, we can expect that EBMT performs better with smaller parallel corpus. Moreover, EBMT can translate in good quality when it has good example match. However, it has low-coverage issues due to low parallel corpus. Another research [4] reported a phrasal EBMT for translating English to Bengali. They did not provide any evaluation of their EBMT. They did not clearly explain their translation generation, specially the word reorder mechanism. One research [5] reported an EBMT for translating news headlines. Another related research [6] proposed EBMT for English-Bengali using WordNet in limited manner.

Chunk parsing was first proposed by Abney [7]. Although EBMT using chunks as the translation unit is not new, it has not been explored widely for low-resource Bengali language yet [8]. used syntactic chunks as translation units for improving insertion or deletion words between two distant languages. However, this approach requires an example base with aligned chunks in both source and target language. In our example-base only source side contains chunks and target side contains corresponding translation string.

Template based approaches increased coverage and quality in previous EBMT. Moreover, [9] showed that templates can still be useful for EBMT with statistical decoders to obtain longer phrasal matches. Manually clustering the words can be a time consuming task. It would be less time consuming to use standard avail-able resources such as WordNet for clustering. That is why in our system, we used <lexical filename> information for each English words, provided by WordNet-Online for clustering the proposed CST.

For low-resource language [10] proposed source language adaptation approaches. Their approach needs large corpus which is similar to the low-resource language. However, Bengali has no such similar language which has large parallel corpus. Moreover, their approach requires a low-resource language as target language and a rich language as a source a language.

3 EBMT System Architecture

The Fig. 1 shows the proposed EBMT architecture. The dotted rectangles identified the main contribution area of this research. During the translation process, at first, the input sentence is parsed into chunks using OpenNLP Chunker. The output of Source Language Analysis step is the English chunks. Then the chunks are matched with the example-base using the Matching algorithm as described in Sect. 4.2. This process provides the CSTs candidates from the example-base and it also identify the unknown words in CSTs. In unknown word translation step, using our proposed mechanism in Sect. 4.3, the system finds translation candidates for the identified

Fig. 1 Proposed EBMT
system architecture

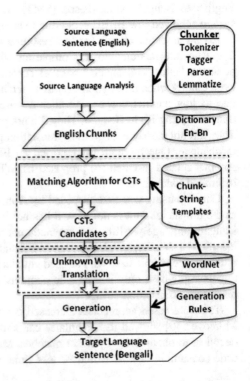

unknown words from WordNet. Then in Generation process WordNet helps to translate determiners and prepositions correctly to improve MT quality [6]. Finally using the generation rules we output the target-language strings. Based on the EBMT system architecture in Fig. 1, we built an English-to-Bengali EBMT system.

4 Chunk-String Templates

In this research we proposed EBMT based on (CST), which is especially useful for developing a MT system for high-resource in source language to low-resource in target language. CST consists of a chunk in the source language (English), a string in the tar-get language (Bengali), and the word alignment information between them. From the English-Bengali aligned parallel corpus CSTs are generated automatically.

Table 1 shows sample word-aligned parallel corpus. Here the alignment information contains English position number for each Bengali word. For example, the first Bengali word "বিশ্বব্যাপী" is aligned with 11. That means "বিশ্বব্যাপী" is aligned with "worldwide", the 11th word in the English sentence. Although the last Bengali word "মাতৃভাষা" is aligned with 4, the word meaning includes "the native language". Therefore, the alignment information does not have 3rd and 5th words.

The example-base of our EBMT is stored as CST. CST consists of <c; s; t>, where c is a chunk in the source language (English), s is a string in the target language (Bengali), and t is the word alignment information between them.

4.1 Generate CSTs

A chunk is a non-recursive syntactic segment which includes a head word with related feature words. In this paper OpenNLP has been used for chunking purpose. For example, "[NP a/DT number/NN]", is a sample chunk. Here NP, DT, NN are parts of speech (POS) Tag defined in Penn Treebank tag set as: proper noun, determiners, singular or mass noun. The third brackets "[]" define the starting and ending of a complete chunk.

Table 1 Example word-aligned parallel corpus

English	Bengali	Align				
Bangla is the native language of 1 2 3 4 5 6 Around 230 million people worldwide 7 8 9 10 11	বিশ্বব্যাপী বাংলা হচ্ছে প্রায় ২৩০ মিলিয়িন মানুষ এর মাতৃভাষা	11 9	1 10	2 6	7 4	8

Fig. 2 Steps of CSTs
generation

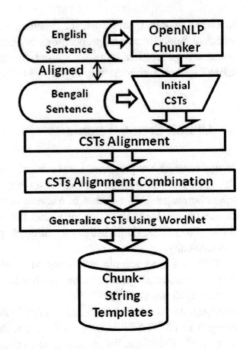

Figure 2 shows the steps of CSTs generation. First the English chunks are auto generated from a given English sentence. Then initial CSTs are generated for each English chunks from the English-Bengali parallel corpus. Each CSTs alignment for all sentences are generated using the parallel corpus. After that the system generate combinations of CSTs. Finally, the system produces CSTs by generalizing using WordNet to achieve wide-coverage.

4.1.1 OpenNLP Chunker

In the first step, using OpenNLP chunker, we prepare chunks of the English sentences from the word aligned English-Bengali parallel corpus.

Input of this step: *"Bangla is the native language of around 230 million people worldwide."*

Output of this step: *"[NP Bangla/NNP] [VP is/VBZ] [NP the/DT native/JJ language/NN] [PP of/IN] [NP around/RB 230/CD million/CD people/NNS] [ADVP worldwide/RB] ./."*

4.1.2 Initial CSTs

In this second step, initial CSTs are generated for each English chunks from the English-Bengali parallel corpus. Table 2 shows the initial CSTs for the word aligned parallel corpus given in Table 1.

Table 2 Example of initial CSTs

CST#	English chunk (C)	Bengali (S)	T	Align	Chunk-start-index
CST1	[NP Bangla/NNP]	বাংলা	1	1	0
CST2	[VP is/VBZ]	হচ্ছে	1	2	1
CST3	[NP the/DT native/JJ language/NN]	মাতৃভাষা	2	4	2
CST4	[PP of/IN]	–এর	1	6	5
CST5	[NP around/RB 230/CD million/CD people/NNS]	প্রায় ২৩০ মিলিয়িন মানুষ	1 2 3 4	7 8 9 10	6
CST6	[ADVP worldwide/RB]	বিশ্বব্যাপী	1	11	10

In Table 2 "CST#" is the CSTs number for reference, "C" is the individual English Chunks. "S" is the corresponding Bengali Words. "T" represents the alignment information inside the chunk. "Align" is the same as "Align" in Table 1. "Chunk-Start-Index" equals to the first word position of the chunk in original sentence, minus one. For example, from Table 1 we get:

$$Align = [around, 230, million, people] = [7, 8, 9, 10]$$

The first word of this chunk is "around", which was in word position 7 of English sentence. Subtracting 1, we get the CST5 chunk-start-index is word position 6. For calculating T, the system subtracts the chunk-start-index from each original word alignment. In the above example, the system subtracts the chunk-start-index 6 from each word alignment. Then we get final alignment, $T = [1, 2, 3, 4]$

4.1.3 CSTs Alignment

CSTs alignment stores the English word order and bengali word original sentence alignment information. So that from the initial CSTs the system can reorder the CSTs in Bengali word order.

In this step, the system generates the word order information from Initial CSTs as given in Table 2. For example, Table 3 shows the word order information produced from Table 2. "English order" represent the original English chunks order and "Bengali order" represents the Bengali chunks order by using CSTs in Table 3. For example, [CST6 CST1 CST2 CST5 CST4 CST3] represents the Bengali sentence "বিশ্বব্যাপী বাংলা হচ্ছে প্রায় ২৩০ মিলিয়িন মানুষ–এর মাতৃভাষা".

Table 3 Example of CSTs alignment

CCST#	English order	Bengali order
CCST1	CST1 CST2 CST3 CST4 CST5 CST6	CST6 CST1 CST2 CST5 CST4 CST3

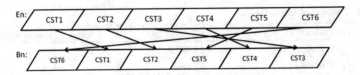

Fig. 3 CSTs alignment

Figure 3 visualize the CSTs alignment from Table 3.

4.1.4 CSTs Alignment Combination

In this step the system generates all possible chunk combinations. The goal is to match source language chunks with as many as possible CSTs. Without CSTs combinations, the system coverage will be low.

From CSTs alignment as given in Table 3, system generates CSTs Combinations. It combines all sequential CSTs. For example, in Fig. 4, circles identified the sequential CSTs combination in Bengali word order. Here CST1 and CST2 can be combined as CCST2, because they are sequential in target language word order.

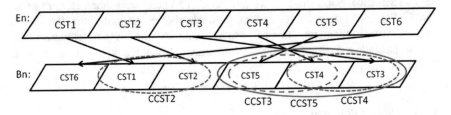

Fig. 4 Chunk alignment

Table 4 CCSTs examples

CCST#	English order	Bengali order
CCST1	CST1 CST2 CST3 CST4 CST5 CST6	CST6 CST1 CST2 CST5 CST4 CST3
CCST2	CST1 CST2	CST1 CST2
CCST3	CST4 CST5	CST5 CST4
CCST4	CST3 CST4	CST4 CST3
CCST5	CST3 CST4 CST5	CST5 CST4 CST3

Table 5 Generalized CSTs

CST#	English chunk (C)	Generalized chunk
CST1	[NP Bangla/NNP]	[NP <noun.communication>/NNP]
CST5	[NP around/RB 230/CD million/CD people/NNS]	[NP around/RB <noun. quantity> people/NNS]

Table 4 contains the whole sentence CCST and the Combined-CSTs (CCSTs) as shown in Fig. 4. The system also produces CSTs combination in source language correspond in target language.

4.1.5 Generalize CSTs Using WordNet

In this step CSTs are generalized by using WordNet to increase the EBMT coverage. To generalize we consider nouns, proper nouns and cardinal number (NN, NNP, CD in OpenNLP tagset) as our first step. For each nouns (NN) or proper nouns (NNP) the system search for the <lexical filename> in WordNet. If the system finds the noun or proper noun, the system replaces that with the <lexical filename> in WordNet. For example, if the system finds "Bangla" in WordNet it replace it with <noun.communication>. For each cardinal number (CD) we simply replace that cardinal number to <noun.quantity>.

Here we show some example of generalized CSTs produced using WordNet in Table 5.

Finally, we get the CSTs database which has three tables: initial CSTs, generalized CSTs and CCSTs. From the example word-aligned parallel sentence of Table 1, system generated 6 initial CSTs, 2 Generalized CSTs and 4 Combined-CSTs.

4.2 Matching Algorithm for CSTs

Matching algorithm for CSTs has three components: search in CSTs, search in CCSTs and selecting CCSTs candidates. The Fig. 5 shows the process of our proposed matching algorithm. The input is the English chunks from the source language sentence. At first the system finds candidate CSTs for each SL chunks from initial CSTs. Search for each chunks using initial CST. Until all chunks are matched the system generalizes the input chunks and search in generalized CSTs. Finally, the system selects best CSTs combination from all the CCSTs candidates.

Fig. 5 Matching algorithm
for CSTs

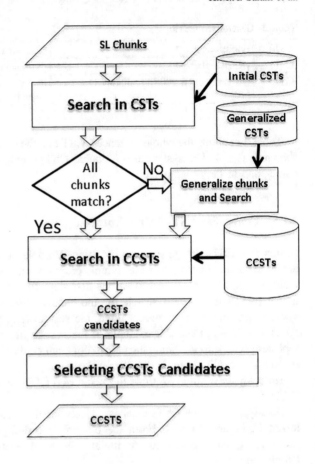

4.2.1 Search in CSTs

To search in CSTs our system first tries to find each chunk in initial CSTs. If it does
not have exact match, it tries to find the linguistically related matches in generalized
CSTs. Linguistically relations are determined by POS tags given in source-language
chunks and the information provided by WordNet. Finally this step provides a set of
matched CSTs. All SL chunks can be matched with at CSTs, generalized CSTs; or
marked as unknown word otherwise.

For example, we have 3 input chunks: [NP English/NNP][VP is/VBZ][NP
the/DT native/JJ language/NN]. Second and third chunks are matched with CST2
and CST3 of initial CSTs in Table 2. But the first chunk [NP English/NNP], has no
match. Then using WordNet the system generalized the input chunk "[NP
English/NNP]" into "[NP <noun.communication>/NNP]". It matched with CST1
of Table 5. This step returns a set of matched CSTs [CST1, CST2, CST3] and
match level (as described in Sect. 4.2.3).

4.2.2   Search in CCSTs

The second step is to search the matched CSTs in CCSTs. The system performs all order CSTs combination search. And it returns CCSTs candidates. For the above example, it returns [CCST1, CCST2, CCST4, CCST5] because these CCSTs include at least one matched CST in [CST1, CST2, CST3]. As this example if more than one CCSTs matches the CSTs, it returns all the CCSTs candidates, to select the best one in the next step.

4.2.3   Selecting CCSTs Candidates

Finally in this step using our selection criteria we choose the suitable CCSTs. From the set of all CCSTs candidates this algorithm selects the most suitable one, according to the following criteria:

1. The more CSTs matched, the better;
2. Linguistically match give priority by following these ranks, higher level is better:

   1. Level 4: Exact match.
   2. Level 3: <lexical filename> of WordNet and POS tags match.
   3. Level 2: <lexical filename> of WordNet match.
   4. Level 1: Only POS tags match.
   5. Level 0: No match found, all unknown words.

4.3   Unknown Word Translation

As in our assumption, the main users of this EBMT will be monolingual people; they cannot read or understand English words written in English alphabet. However, with related word translation using WordNet and Transliteration can give them some clues to understand the sentence meaning. As Bangla language accepts foreign words, transliterating an English word into Bangla alphabet, makes that a Bangla foreign word. For example, in Bangla there exist many words, which speakers can identify as foreign words.

Figure 6 shows the unknown words translation process in a flow chart. Proposed system first tries to find semantically related English words from WordNet for the unknown words. From these related words, we rank the translation candidates using WSD technique and English-Bangla dictionary. If no Bangla translation exists, the system uses IPA-based-transliteration. For proper nouns, the system uses transliteration mechanism provided by Akkhor Bangla Software.

78 K.M.A. Salam et al.

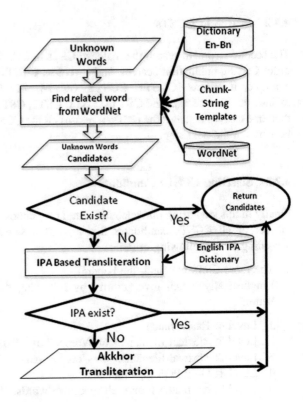

Fig. 6 Steps of handling unknown words

4.3.1 Find Sublexical Translations

For sublexical matching our system divide the unknown word into sublexical units and then find possible translation candidates from these sublexical units. For this the system use following steps:

1. Find the possible sublexical units of the unknown word. For example, the unknown word "bluebird" gets divided into "blue" and "bird".
2. Extract sublexical translations and restrain translation choices.
3. Remove less probable sublexical translations
4. Output translation candidates with the POS tags for the sublexical units of the unknown word.

From the set of all CSTs we select the most suitable one, according to the following criteria:

1. The more exact CSTs matched, the better;
2. Linguistically match give priority by following these ranks, higher level is better:

- Level 4: Exact match.

- Level 3: Sublexical unit match, <lexical filename> of WordNet and POS tags match
- Level 2: Sublexical unit match, <lexical filename> of WordNet match
- Level 1: Only POS tags match.
- Level 0: No match found, all unknown words.

4.3.2 Find Candidates from WordNet

Due to small English-Bangla parallel corpus availability, there is high probability for the MT system to handle unknown words. Therefore, it is important to have a good method for translating unknown words. When the word has no match in the CSTs, it tries to translate using English WordNet and bilingual dictionary for English-Bangla.

Input of this step is unknown words. For example, "canine" is a unknown word in our system. Output of this process is the related unknown words translation.

Find Candidates from WordNet Synonyms

The system first finds the synonyms for the unknown word from the WordNet synsets. Each synset member becomes the candidate for the unknown word. WordNet provide related word for nouns, proper nouns, verbs, adjectives and adverbs. Synonymy is WordNet's basic relation, because WordNet uses sets of synonyms (synsets) to represent word senses. Synonymy is a symmetric relation between word forms. We can also use Entailment relations between verbs available in WordNet to find unknown word candidate synonyms.

Find Candidates Using Antonyms

WordNet provide related word for nouns, proper Antonymy (opposing-name) is also a symmetric semantic relation between word forms, especially important in organizing the meanings of adjectives and adverbs. For some unknown we can get the antonyms from WordNet. If the antonym exists in the dictionary we can use the negation of that word to translate the unknown word. For example, "unfriendly" can be translated as "not friendly". In Bengali to negate such a word we can simply add "না" (na) at the end of the word. So, "unfriendly" can be translated as "বন্ধুত্বপূর্ণ না" (bondhuttopurno na). It helps to translate unknown words like "unfriendly", which improves the (MT) quality.

Hyponymy (sub-name) and its inverse, hypernymy (super-name), are transitive relations between synsets. Because there is usually only one hypernym, this semantic relation organizes the meanings of nouns into a hierarchical structure. We need to process the hypernyms to translate the unknown word.

Find Candidates Using Hypernyms

For nouns and verbs WordNet provide hypernyms, which is defined as follows:

Y is a hypernym of X if every X is a (kind of) Y.

For example "canine" is a hypernym of noun "carnivore", because every dog is a member of the larger category of canines. Verb example, "to perceive" is an hypernym of "to listen". However, WordNet only provides hypernym(s) of a synset, not the hypernym tree itself. As hypernyms can express the meaning, we can translate the hypernym of the unknown word. To do that, until any hypernym's Bangla translation found in the English-Bangla dictionary, we keep discovering upper level of hypernym's. Because, nouns and verbs are organized into hierarchies, defined by hypernyms or is-a-relationships in WordNet. So, we considered lower level synset words are generally more suitable then the higher level synset words.

This process discovers the hypernym tree from WordNet in step by step. For example, from the hypernym tree of dog, we only had the "animal" entry in our English-Bengali dictionary. Our system discovered the hypernym tree of "dog" from WordNet until "animal". Following is the discovered hypernym tree:

```
dog, domestic dog, Canis familiaris
=> canine, canid
  => carnivore
    => placental, placental mammal, eutherian mammal
      => mammal
        => vertebrate, craniate
        => chordate
          => animal => ...
```

This process search in English-Bangla dictionary, for each of the entry of this hypernym tree. So at first we used the IPA representation of the English word from our dictionary, then using transliterating that into Bengali. Then system produce "a kind of X"—এক·ধরনের·X [ek dhoroner X]. For the example of "canine" we only had the Bengali dictionary entry for "animal" from the whole hypernym tree. We translated "canine" as the translation of "canine, a kind of animal", in Bangla which is "ক্যানাইন, এক ধরনের পশু" [kjanain, ek dhoroner poshu].

Similarly, for adjectives we try to find "similar to" words from WordNet. And for Adverbs we try to find "root adjectives".

Finally, this step returns unknown words candidates from WordNet which exist in English-Bangla dictionary.

Using the same technique described above, we can use Troponyms and Meronyms to translate unknown words. Troponymy (manner-name) is for verbs what hyponymy is for nouns, although the resulting hierarchies are much shallower. Meronymy (part-name) and its inverse, holonymy (whole-name), are complex semantic relations. WordNet distinguishes component parts, substantive parts, and member parts.

4.4 Rank Candidates

To choose among the candidates for the unknown word, we need to rank all the candidates. Especially polysemous unknown words need to select the adequate WordNet synset to choose the right candidate. The system performs Google search with the input sentence as a query, by replacing the unknown word with each candidate words. We add quotation marks in the input sentence to perform phrase searches in Google, to find the number of in documents the sentence appear together. If the input sentence with quotation mark returns less than 10 results, we perform Google search with four and two neighbor chunks. Finally, the system ranks the candidate words using the Google search hits information.

For example, the input sentence in SL is: This dog is really cool. The system first adds double quotation with the input sentence: "This dog is really cool", which returns 37,300 results in Google. Then the system replaces the unknown word "dog" from discovered hypernym tree. Only for "This animal is really cool.", returned 1560 results by Google. That is why "animal" is the second most suitable candidate for "dog". However, other options "This domestic dog is really cool." or "This canine is really cool." etc. returns no results or less than 10 results in Google. So in this case we search with neighbour chunks only. For example, in Google we search with:

"This mammal is" returns 527,000 results;
"This canid is" returns 503,000 results;
"This canine is" returns 110,000 results;
"This carnivore is" returns 58,600 results;
"This vertebrate is" returns 2460 results;
"This placental is" returns 46 results;
"This craniate is" returns 27 results;
"This chordate is" returns 27 results;
"This placental mammal is" returns 6 result;

Finally, the system returns the unknown word candidates: mammal, canid, canine, carnivore, vertebrate, placental, craniates, chordate, placental mammal.

4.4.1 Final Candidate Generation

In this step, we choose one translation candidate. If any of the synonyms or candidate word exist in English-Bangla dictionary, the system translates the unknown word with that synonym meaning. If multiple synonyms exist, then the entry with highest Google search hits get selected. English-Bangla dictionary also contains multiple entries in target language. For WSD analysis in target language, we perform Google search with the produced translation by the system. The system chooses the entry with highest Google hits as final translation of the unknown word.

For example, for unknown word "dog", animal get selected in our system. However, if there were no candidates, we use IPA-Based-Transliteration.

4.4.2 IPA-Based-Transliteration

When unknown word is not even found in WordNet, we use IPA-Based transliteration using the English IPA Dictionary [11]. Output for this step is the Bangla word transliterated from the IPA of the English word. In this step, we use English-Bangla Transliteration map to transliterate the IPA into Bangla alphabet. From English IPA dictionary the system can obtain the English words pronunciations in IPA format. Output for this step is the Bengali word transliterated from the IPA of the English word. In this step, we use following English-Bengali Transliteration map to transliterate the IPA into Bengali alphabet. Figure 7 shows our proposed English-Bengali IPA chart for vowels, diphthongs and consonants. Using rule-base we transliterate the English IPA into Bangla alphabets. The above IPA charts leaves out many IPA as we are considering about translating from English only. To translate from other language such as Japanese to Bangla we need to create Japanese specific IPA transliteration chart. Using the above English-Bangla IPA chart we produced transliteration from the English IPA dictionary. For examples: pan(pæn): প্যান; ban(bæn): ব্যান; might(maIt): মাইট.

However, when unknown word is not even found in the English IPA dictionary, we use transliteration mechanism of Akkhor Bangla Software as given in Fig. 8. For example, for the word "Muhammod" which is a popular Bangla name, Akkhor transliterated into "মুহাম্মদ" in Bangla.

4.5 Translation Generation

In this EBMT architecture we used Rule-Based generation method. Using dictionary and WordNet rules, we can accurately translate the determiners in Bengali. For translating determiner, we adapted [6] proposals to use WordNet.

To reorder the CSTs for partial match in CCSTs, we remove the unmatched CSTs. Based on the morphological rules we change the expression of the words.

Here WordNet provided required information to translate polysemous determiners accurately. The system compared with the <lexical filename> of WordNet for the word NN. If the word NN is "<noun.person>", then determiner "a" will be translated as "ekjon". Otherwise "a" will be translated as "ekti".

For example "a boy" should be translated to "ekti chele" as boy is a person. "ekkhana chele" is a wrong translation, because "ekkhana" can be used only for NNs which is not a person.

For Bengali word formation we have created morphological generation rules especially for verbs. These rules are constructed by human.

Mouth narrower vertically	[iː] ই / ি sleep /sliːp/	[I] ই / ি slip /slIp/	[ʊ] উ / ু book /bʊk/	[uː]উ/ ু boot /buːt/
	[e] এ / ে ten /ten/	[ə] আ / আ after /aːftə/	[ɜː] আ / আ bird /bɜːd/	[ɔː] র bored /bɔːd/
Mouth wider vertically	[æ]এ্যা/ ্যা cat /kæt/	[ʌ] আ/আ cup / kʌp/	[ɑː] আ/ আ car / cɑːr/	[ɒ] অ hot /hɒt/

English-Bengali IPA mapping for vowels

[Iə] ইয়া/িয়া beer /bIər/	[eI] এই/ েই say /seI/	
[ʊə] উয়া/ ুয়া fewer /fjʊər/	[ɔI] অয়/য় boy /bɔI/	[əʊ] ও / ো no /nəʊ/
eə ঈয়া/ ীয়া bear /beər/	[aI] ইি / আই high /haI/	[aʊ] আউ /িউ cow /kaʊ/

English-Bengali IPA mapping for diphthongs

[p] প pan /pæn/	[b] ব ban /bæn/	[t] ট tan /tæn/	[d] ড day /deI/	[ʧ] চ chat /ʧæt/	[ʤ] জ judge /ʤʌʤ/	[k] ক key /kiː/	[g] গ get /get/
[f] ফ fan /fæn/	[v] ভ van / væn/	[θ] থ thin /θIn/	[ð] দ than /ðæn/	[s] স sip /sIp/	[z] জ zip / zIp/	[ʃ] শ ship /ʃIp/	[ʒ] স vision /vIʒ^n/
[m] ম might /maIt/	[n] ন night /naIt/	[ŋ]ং/ঙ thing /θIŋ/	[h] হ height /haIt/	[l] ল light /laIt/	[r] র right /raIt/	[w] ৱ white/hwaIt/	[j]ইয়ে yes /jes/

English-Bengali IPA mapping for consonants

Fig. 7 English-Bengali IPA mapping

বাংলা	অ	আ	ই	ঈ	উ	ঊ	ঋ	এ	ঐ	ও	ঔ
English	A	a/aa/a	i/i	I/ee/'I	u/u	U/U	ri/ri	e/e	oi/'oi	o/o	ou/ou

বাংলা	ক	খ	গ	ঘ	ঙ	চ	ছ	জ	ঝ	ঞ
English	k	kh	g	gh	Ng	ch	Ch	j	jh	Y
বাংলা	ত	থ	দ	ধ	ন	ট	ঠ	ড	ঢ	ণ
English	t	th	d	dh	n	T	Th	D	Dh	N
বাংলা	প	ফ	ব	ভ	ম	য	র	ল	শ	ষ
English	p	f/ph	b	bh/v	m	z	r	l	sh	S
বাংলা	স	ক্ষ	হ	ড়	ঢ়	য়	ৎ	ঃ	ঁ	০
English	S	k-S	h	R	rh	y	ng	:	~	
বাংলা	১	২	৩	৪	৫	৬	৭	৮	৯	০
English	1	2	3	4	5	6	7	8	9	1
বাংলা	কা	কে	কি	কু	কো	ক্র	ক্রে	ক্রি	কৃ	কৃ
English	ka	ke	ki	ku	kO	kro	kre	kre e	kru	krU
বাংলা	কী	চী	মী	কূ	মৃ	বৃ	পৃ	নৃ	ক্য	ব্য
English	kI	chI	mI	kU	mU	bU	NU	nU	k-z	b-z

Fig. 8 Akkhor phonetic mapping for bengali alphabets

5 Evaluation

We did wide-coverage and quality evaluations for the proposed EBMT with CSTs, by comparing with baseline EBMT system. Wide-coverage evaluation measures the increase of translation coverage. Quality evaluation measures the translation quality through human evaluation. Because of the unavailability of large parallel corpus for English-Bengali language pair, we could not evaluate the BLEU score.

Baseline system architecture has the same components as described in Fig. 1, except for the components inside dotted rectangles. Matching algorithm of baseline system is that not only match with exact translation examples, but it can also match with POS tags. The Baseline EBMT use the same training data: English-Bengali parallel corpus and dictionary, but does not use CSTs, WordNet and unknown words translation solutions. Currently from the training data set of 2000 word aligned English-Bengali parallel corpus, system generated 15,356 initial CSTs, 543 Generalized CSTs and 12,458 Combined-CSTs.

The development environment was in windows using C Sharp language. Our test-set contained 336 sentences, which are not same as training data. The test-set includes simple and complex sentences, representing various grammatical phenomena. Table 6 shows the distribution of sentences in different categories. We have around 20,000 English-Bengali dictionary entries.

Table 6 Different category of test-set sentences

Sentence type	Number of sentences
Simple	136
Complex (Wh-Clause)	50
Complex (Infinitive Clause)	50
Unknown words	100
Total	336

Table 7 Wide-coverage comparison

System modules	Wide-coverage (%)
Baseline EBMT	23
Proposed EBMT with CSTs	80

5.1 Wide-Coverage Evaluation

We calculated the rate of generalized CSTs usage to evaluate the achievement of wide-coverage. To match the English input chunks, baseline EBMT use translation examples and POS matching mechanism from the training data. On the other hand, proposed EBMT use CSTs to match the English input chunks. Table 7 shows the contribution of CSTs to achieve wide-coverage. Here wide-coverage = No. of Matched English chunks/No. of all English chunks in test-set. CSTs improved the wide-coverage by 57 points.

5.2 Quality Evaluation

5.2.1 CSTs Evaluation

Quality evaluation measures the translation quality through human evaluation. Table 8 shows the human evaluation of the proposed EBMT system with CSTs only.

Table 8 Human evaluation using same test-set

Translation quality	Grade	EBMT + CSTs	Google translate
Perfect translation	A	25.60	19.00
Good translation	B	38.69	30.00
Medium translation	C	19.64	27.00
Poor translation	D	16.07	24.00
Total		100%	100%

Table 9 Human evaluation quality explanation

Translation quality	Word selection	Word order	Functional word usage	Example translation produced
Perfect translation	YES	YES	YES	জাপানিজ হচ্ছে প্রায় ১২০ মলিয়িন মানুষ–এর মাতৃভাষা
Good translation	YES	YES	NO	প্রস্তুতি ম্যাচ সেঞ্চুরিটা কাজ লাগল ইমরুল কায়েসের
Medium translation	YES	NO	YES/NO	ইমরুল ফরম দেওয়া ফিরিছেনে বাংলাদেশে ফেরার ইঙ্গিত
Poor translation	NO	NO	NO	Roseland ওয়েস্ট Pullman, এবং Riverdale আট ক্যাথলকি প্যারিশ সমন্বয়ে

Table 9 shows the explanation of translation quality used in our human evaluation process. Word selection means whether the system could choose a correct word candidate. Word order measures whether the words position in the translated sentence is grammatically correct. Functional word usage means whether the system could choose a correct functional word. Considering these quality elements, we have evaluated the translation quality.

Perfect Translation means there is no problem in the target sentence. Good Translation means the target sentence is not grammatically correct because of wrong functional word, but still understandable for human. For example, in Table 9 we graded the example translation produced in Good Translation category because the words ম্যাচ and কাজ is not in correct functional word form but still the sentence is understandable by human. Medium means there are several mistakes in the target sentence, like wrong functional word and wrong word order. In the example sentence we can see that the word order doesn't make any sense even though the word selection is correct. So human cannot understand the translated sentences in medium category. Poor Translation means there are major problems in the target sentence, like non-translated words, wrong word choice and wrong word order. In the example, the produced translation does not make sense due to wrong word selection with wrong word order.

Only perfect and good translations were "acceptable". Because even though the system chooses the correct word without generating the correct word order the translated sentence will be grammatically incorrect and may not be understandable.

Currently 64.29% of the test-set translations produced by the system were acceptable, produced by the system with proposed CSTs only. Around 48.81 points of poor translation produced by EBMT Baseline was improved using the proposed system with CSTs.

The identified main reasons for improving the translation quality is our solution using CSTs generalization and sub-sentential match. Because of these contributions of CSTs some test-set sentence improved from "poor" or "medium" translation to "acceptable" translation.

Table 10 Comparison of CSTs output with google translate using same test set

#	English	EBMT + CSTs	Google translate
1.	Japanese is the native language of around 120 million people	জাপানিজ হচ্ছে প্রায় ১২০ মলিয়িন মানুষ–এর মাতৃভাষা (A)	জাপানি প্রায় 120 কোটি মানুষের মাতৃভাষা (C)
2.	The name Bangladesh was originally written as two words, Bangladesh	বাংলাদেশে নামটি মূলত দুই শব্দে লেখা হতো, বাংলা দেশ (A)	নাম বাংলাদেশে মূলত দুই বোথ ওয়ার্ল্ডস, বাংলাদেশে হিসিবে লেখা হয়ছে (D)
3.	After high school, Obama moved to Los Angeles in 1979 to attend occidental college	হাই স্কুলের পরে, ওবামা লস এঞ্জলেসে এ ১৯৭৯ সালে অকসডিনেটাল কলেজে যায় (A)	উচ্চ বিদ্যালয় পরে, ওবামা অকসডিনেটাল কলেজে যোগ দিতে 1979 সালে লস এঞ্জলেসে সরানো. (C)

Table 10 shows some good example translations comparison between EBMT + CSTs and Google Translate. It also shows the translation quality in bracket (A, B, C, D: Perfect, Good, Medium, Poor). In #1 translation, Google translate mistranslated the word "million" to "crore". Google translate could not translate #2 properly with wrong word choices and wrong word orders which results a poor translation quality. In #3 translation, it shows Google never translation the digits into Bangla digits and the word orders are not correct like other complex sentences. On the other hand, CSTs performed well in all 3 examples, which demonstrate the goodness of using CSTs for low resource language.

We observed some drawbacks of using CSTs with generalization using WordNet as well. Sometimes our system chooses the wrong synset from the WordNet. As a result, some test-set still produced "poor" translation.

5.2.2 Unknown Words Evaluation

We also did quality evaluation for our unknown words solution. Table 11 shows the human evaluation of the EBMT system with CSTs and unknown word solution. Currently 67.85% of the test-set translations were acceptable, produced by the system with proposed CSTs and unknown words solutions. Comparing with EBMT + CSTs, unknown words mechanism improved translation quality by 3.56 points in human evaluation. We also compare our system with Google translate which is the most popular MT system for English-Bengali language pair.

Table 11 Human evaluation of unknown words using same test-set

Translation quality	Grade	EBMT + CSTs + Unknown words	Google translate
Perfect translation	A	30.95	19.00
Good translation	B	36.90	30.00
Medium translation	C	18.75	27.00
Poor translation	D	13.39	24.00
Total		100.00	100.00

Table 12 Human evaluation of unknown words using same testset

#	English	EBMT + CSTs + Unknown words	Google translate
1.	Are you looking for an aardvark?	আপনি কি আর্ডভার্ক, এক ধরনের পশু খুঁজছেন?(A)	আপনি একটি aardvark খুঁজছেন? (C)
2.	This dog is really cool.	ডগ, এক ধরনের পশু আসলেই দারুন (A)	এই কুকুর সত্যিই শীতল হয়. (C)
3.	WordNet is a lexical database for the English language	শব্দজাল হচ্ছে ইংরেজি ভাষার জন্য একটা আভিধানিক ডাটাবেস (A)	WordNet ইংরেজি ভাষার জন্য একটা আভিধানিক ডাটাবেস (B)
4.	Sublexical units of a word are selected in parallel and are subsequently ordered	শব্দের উপ-আভিধানিক অংশ নির্বাচন করা হয় সমান্তরাল ভাবে এবং অনুক্রম অনুসারে (A)	একটা শব্দের sublexical ইউনিট সমান্তরাল মধ্যে নির্বাচন করা হয় এবং পরবর্তীকালে আদেশে করা হয়(D)
5.	The bluebird are a group of medium-sized, mostly insectivorous or omnivorous bird in the world	নীলপাখি হচ্ছে এক ধরণের পাখি যা বিশ্বের মাঝারি আকারের, সাধারণত কীটভক্ষক এবং সর্বভুক পাখির একটা গ্রুপ (A)	bluebirds বিশ্বের মাঝারি আকারের, বেশির ভাগ কীটভক্ষক এবং সর্বভুক পাখি একটা গ্রুপ (D)

As we used same test-set, the result of Google MT is same for both Table 8 and 11. Our EBMT could translate better than Google because of our novel CSTs and unknown words translation mechanism.

Table 12 shows sample translation examples produced by EBMT + CSTS with unknown words solution compared with Google translate. It also shows the translation quality in bracket (A, B, C, D: Perfect, Good, Medium, Poor).

As "aardvark" and "dog" are unknown words, Google translate produced medium translation for #1 and #2. As a result the translation quality improved to "good" quality. All these examples demonstrate the effectiveness of our proposed solution for translating unknown words.

6 Discussion

6.1 Wide-Coverage of Adequate Determiner Evaluation

As we used WordNet to translate using adequate determiner, we measured the increase of translation coverage as following.

$$\text{wide-coverage} = \frac{\text{No. of system generate adequate determiner}}{\begin{array}{c}\text{No. of all adequate determiner}\\ \text{(from example Human evaluation sentences)}\end{array}}$$

Table 13 Wide-coverage comparison

System modules	Wide-coverage (%)
Baseline EBMT	24
Proposed EBMT with WordNet	65

Table 13 shows the EBMT system performance improvement for the test data of 336 sentences. In these test sentences we had 107 adequate determiners. The baseline EBMT produced 34 adequate determiners, which is 24% of all adequate determiners. The proposed EBMT produced 93 adequate determiners, which is 65% of all adequate determiners. Our proposed EBMT system improved the wide-coverage of adequate determiners by 41 points. We found generalized CSTs are also effective for achieving wide-coverage in translating determiners.

6.2 Grammatical Structures of Test-Set Sentences

English sentences in our test-set can be classified in four types: Declarative, Imperative, Interrogative and Exclamatory sentences. These sentences can also be classified using following complexity types: Simple, Compound, Complex and Compound-Complex. Current EBMT system performance depend on the quality of English chunker.

7 Conclusion and Future Works

We proposed to use ontology to improve the quality of EBMT system for low-resource language. Our EBMT system is effective for low resource language like Bengali. We used WordNet to translate the unknown words which are not directly available in the dictionary. To translate an English sentence, it is first parsed into chunks. Then the chunks matched with the CSTs to find translation candidates. Then the system determines translation candidates for the identified unknown words from WordNet. Finally using generation rules the target-language strings has been produced.

Using this method, our proposed EBMT system improved the wide-coverage by 57 points and quality by 48.81 points in human evaluation. Currently 64.29% of the test-set translations by the system were acceptable. Because our system can generate more general CSTs, and it increases the quality for low-resource language. Unknown words mechanism improved translation quality by 3.56 points in human evaluation. Currently 67.85% of the test-set translations were acceptable, produced by the system with proposed CSTs and unknown words solutions.

In future we would like to use statistical language model to improve the generation quality. We would like to evaluate the system with other high resource to low-resource language pair.

References

1. Miller, G.A.: WordNet: a lexical database for english. Commun. ACM **38**(11), 39–41 (1995)
2. Dasgupta, S., Wasif, A., Azam, S.: An optimal way towards machine translation from english to bengali. In: Proceedings of the 7th International Conference on Computer and Information Technology (ICCIT) (2004)
3. Islam, M.Z., Tiedemann, J., Eisele, A.: English to bangla phrase-based machine translation. In: Proceedings of the 14th Annual conference of the European Association for Machine Translation (2010)
4. Naskar, S.K., Bandyopadhyay, S.: Handling of prepositions in english to bengali machine translation. In: The proceedings of Third ACL-SIGSEM Workshop on Prepositions, EACL, Trento, Italy (2006)
5. Saha, D.: Sudip Kumar Naskar, Sivaji Bandyopadhyay. A Semantics-based English-Bengali EBMT System for translating News Headlines, MT Xummit X (2005)
6. Khan, M., Salam, A., Khan, M., Nishino, T.: Example Based English-Bengali Machine Translation Using WordNet. TriSAI, Tokyo (2009)
7. Abney, S.: Parsing by Chunks, pp. 257–278. Kluwer Academic Publishers, In Principle—Based Parsing (1991)
8. Kim, J.D., Brown, R.D., Carbonell, J.G.: Chunk-Based EBMT. EAMT, St Raphael, France (2010)
9. Gangadharaiah, R., Brown, R.D., Carbonell, J.G.: Phrasal equivalence classes for generalized corpus-based machine translation. In: Gelbukh, A. (ed.) Computational Linguistics and Intelligent Text Processing. Lecture Notes in Computer Science, vol. 6609, pp. 13–28. Springer, Berlin/Heidelberg (2011)
10. Wang, P., Nakov, P., Ng, H.T.: Source language adaptation approaches for resource-poor machine translation. *Comput. Linguist.* **42**, 277–306, 2 June(2016)
11. Khan, M., Salam, A., Yamada, S., Nishino, T.: Example-Based Machine Translation for Low-Resource Language Using Chunk-String Templates, *13th Machine Translation Summit*. Xiamen, China (2011)

A Fast Area Labeling Method Using Auxiliary Lines

Noboru Abe, Kohei Kuroda, Yosuke Kamata and Shogo Midoritani

Abstract This paper considers placing area labels on a map such as labels for cities, prefectures, and lakes. We propose a real-time method to determine label positions using the intersections of the auxiliary and boundary lines of a given area. Experimental results demonstrate that the proposed method is capable of real-time processing and can determine effective label positions.

Keywords Label placement · Real-time processing · Area labeling · Auxiliary line

1 Introduction

Let A be an area on a map such as a city, prefecture, or lake. Each area A has a label with a specified size. In this study, we consider placing a label that corresponds to a name or attribute inside each area A. Placing labels for graphical features appropriately is important in geographical information systems and cartography. Placing an area label on an edge or in a narrow space of A should be avoided. In addition, placing the entire area label outside A should also be avoided; however, it may occur that part of the label is placed outside when A is very small.

In recent years tablet devices are rapidly spreading. The tablet devices can easily enlarge or reduce the map. In such a case, it is important to determine the label position quickly. For industrial applications, an area labeling method should satisfy the following conditions (personal communication, Dawn Corporation employees).

- The labeling method is capable of real-time processing on tablet devices, i.e., it can define all label positions of areas in the display frame in 100 ms.
- The labeling method positions labels (including labels for partially displayed areas) effectively, e.g., near the centroid of large convex polygon in the area.

N. Abe (✉) · K. Kuroda · Y. Kamata · S. Midoritani
Faculty of Information and Communication Engineering, Osaka Electro-Communication
University, Neyagawa-Shi, Osaka 572–8530, Japan
e-mail: abe@osakac.ac.jp

© Springer International Publishing AG 2018
R. Lee (ed.), *Applied Computing & Information Technology*,
Studies in Computational Intelligence 727, DOI 10.1007/978-3-319-64051-8_6

Note that overlapping labels should also be avoided, though the overlaps can be eliminated by enlarging the map on the tablet device.

Many methods have been proposed for point labeling [1–7]. Kameda and Imai [6] and Abe et al. [7] presented methods for both point and line labeling. By contrast, to date, area labeling has received little attention [8]. Aonuma et al. [9] and Dorschlag et al. [10] proposed methods that employ Voronoi diagrams. However, these methods are not capable of real-time processing if A has a significant number of boundary lines. In such a case, they require significant time to determine label positions because they induce Voronoi diagrams from all endpoints of the boundary lines of A. Kakoulis and Tollis [3] and Wagner et al. [4] proposed methods to place area, point, and line labels. Both methods first create several candidate label positions for each area, point, and line. Hereafter, we refer to such candidates as *label candidates*. In addition, both methods require appropriate label candidates to achieve effective label placement; however, a process to create such label candidates for given areas has not been described in detail. Rylov et al. [8] proposed a method to place labels outside the area, which differs from our objective.

In this paper, we propose a real-time label placement method. The proposed method can place a label near the centroid of a fairly large convex polygon in the area. The proposed method employs the intersections of auxiliary horizontal and vertical lines and area boundary lines to determine an appropriate label position.

The remainder of this paper is organized as follows. In Sect. 2, we explain the proposed method. In Sect. 3, we show experimental results. In Sect. 4, we consider how to apply the proposed method when only part of a target area is displayed. We also consider how to handle inclusion relationships such as cities within prefectures. Conclusions are presented in Sect. 5.

2 Proposed Method

In this section, we describe a simple and fast area labeling method and its limitations. Then, we explain the proposed method and preprocessing steps employed to increase its speed. After that, we describe the algorithm of the proposed method.

2.1 Simple and Fast Method

Positioning labels at the center of the minimum bounding rectangle of each A is simple and fast. Such a method frequently achieves good label placement for an area that forms a convex polygon (Fig. 1a). However, this method often places the label in a narrow section of an area that forms a concave polygon (Fig. 1b) and can place the label outside the area (Fig. 1c).

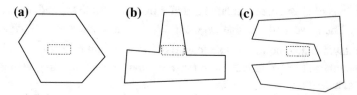

Fig. 1 Label placement using the simple method (*solid lines, area boundary lines, dashed rectangles*, labels)

2.2 Proposed Label Placement Method

The proposed area labeling method employs the distances between the intersections of some auxiliary lines and the boundary lines of A. Note that auxiliary lines can be horizontal or vertical.

Assume that all of area A is displayed. The proposed method first finds the minimum bounding rectangle R of A, as shown in Fig. 2. This can be achieved quite quickly by employing a preprocessing step that stores the left-, right-, top-, and bottom-most endpoints of the boundary lines of A. In Fig. 2, the auxiliary line h (dashed-dotted) bisects R horizontally, and ch_1, ch_2, ch_3, and ch_4 denote the intersections of auxiliary line h and the boundary lines of A. Here let k be the number of intersections and i be an odd number less than k. Note that k is always an even number, and each line segment $\overline{ch_i ch_{i+1}}$ exists inside A. Then, we find the maximum length of $\overline{ch_i ch_{i+1}}$ from each i and let m be such i. Note that in Fig. 2 $m = 1$.

In Fig. 3, the vertical line v' (dashed double-dotted) is derived from $\overline{ch_m ch_{m+1}}$ and is drawn through the midpoint of $\overline{ch_m ch_{m+1}}$. Let cv'_1, cv'_1, ... be the

Fig. 2 Intersections of auxiliary line h and boundary lines of A

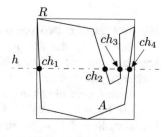

Fig. 3 Label candidate determined by the proposed method

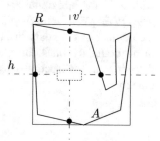

intersections of v' and the boundary lines of A from bottom to top and let j be an odd number. Then, we find the line segment $\overline{cv'_j cv'_{j+1}}$ that intersects $\overline{ch_m ch_{m+1}}$. We create a label candidate (dashed rectangle) at the midpoint of $\overline{cv'_j cv'_{j+1}}$. Additional label candidates can be derived from other auxiliary lines (including vertical ones) in the same manner.

The proposed method selects the final label position from six label candidates. Three candidates are derived from three horizontal auxiliary lines that quadrisect R and remaining three are derived from three vertical lines that quadrisect R.

Here let hor_{lc} and $vert_{lc}$ be the horizontal and vertical lines through the midpoint of each label candidate lc, respectively. In addition, let L_{lc} be a set of intersections of hor_{lc} and the boundary lines of A that exist at the left side of the midpoint of lc. Similarly, let R_{lc}, A_{lc}, and B_{lc} be sets of intersections to the right of, above, and below the midpoint of lc, respectively. To select the final label position, the proposed method computes the following distances for each label candidate lc.

$dist_l$ Distance between the midpoint of lc and the right-most intersection of L_{lc}.
$dist_r$ Distance between the midpoint of lc and the left-most intersection of R_{lc}.
$dist_a$ Distance between the midpoint of lc and the bottom-most intersection of A_{lc}.
$dist_b$ Distance between the midpoint of lc and the top-most intersection of B_{lc}.

In addition, the ratios of $dist_l$ and $dist_r$ to the width of the label, and the ratios of $dist_a$ and $dist_b$ to the height of the label are computed. We define the minimum value of these four ratios as the flexibility of the label candidate. In the proposed method, the label candidate with maximum flexibility is selected as the final label position.

2.3   Preprocessing Steps

Here we explain two preprocessing steps implemented for quick detection of intersections. In the first preprocessing step, we compute the midpoints of the boundary lines of A. The midpoints are stored in increasing order relative to the horizontal and vertical directions. By executing a binary search on these two progressions, we can obtain one intersection quickly if intersections of an auxiliary line and the boundary lines of A exist. In the second preprocessing step, for each boundary line bl of A, we store all boundary lines of A that have a part common with bl in the horizontal and vertical directions. For example, for $\overline{p_1 p_2}$ in Fig. 4, $\overline{p_7 p_8}$, $\overline{p_8 p_9}$, $\overline{p_{12} p_{13}}$, and $\overline{p_{13} p_{14}}$ are stored for the horizontal direction. In most cases, the number of boundary lines to be stored is significantly smaller than the number of all boundary lines. By considering only the stored boundary lines of A after detecting an intersection by binary search, we can detect all intersections quite quickly.

Fig. 4 *Boundary lines* to be
stored for the *horizontal*
direction

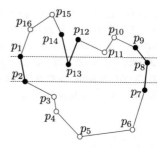

2.4 Algorithm Description

The algorithm of the proposed method is described as follows:

(1) Execute the following preprocessing steps.
   (1-a) Store the left-, right-, top-, and bottom-most endpoints of the boundary lines of A.
   (1-b) Store the midpoints of the boundary lines of A in increasing order relative to the horizontal and vertical directions.
   (1-c) For each boundary line bl of A, store all boundary lines of A that have a part common with bl in the horizontal and vertical directions.
(2) For each auxiliary line, execute the following steps.
   (2-a) Obtain one intersection of the auxiliary line and the boundary lines of A by binary search.
   (2-b) Detect all intersections by considering only the stored boundary lines by step (1-c).
   (2-c) Create a label candidate.
(3) Select the label candidate with maximum flexibility as the final label position.

3 Computational Experiments

We performed computational experiments to confirm the effectiveness of the proposed method. In this section, we explain the experimental conditions, the evaluation method, and the experimental results.

3.1 Experimental Conditions

As the input, we used ten areas obtained from Fundamental Geospatial Data provided by the Geospatial Information Authority of Japan [11]. Each area has indents or narrow parts; therefore, they were not suitable for label placement using the simple method such as described in Sect. 2.1. Label height was set to 5% of the height of the corresponding area. Label width was set to the height times $let/2$, where let is the number of letters in the label. We used an Intel Core i7-6700 processor. Note that we set the clock speed to 1.0 GHz to examine the behavior on a low performance computer such as a tablet. The experiments were implemented using the C programming language.

3.2 Evaluation

To evaluate the performance of the proposed method, we employed a method to enumerate the convex polygons inside area A. Note that this requires significant computation time.

The inputs of this method are the coordinates of the boundary lines of area A, label size, and the coordinates of the center of the label position obtained by the proposed method. First, label width is increased or decreased such that the label becomes a square. Note that the coordinates of boundary lines and the center of the label position also change. Here let A' denote an increased or reduced area A.

Next, starting from an endpoint of a boundary line, the method considers at the other endpoints in clockwise order and connects them by a line segment. Assume that the starting endpoint is denoted p_a and all endpoints $p_1, p_2, p_3, \cdots, p_{a-1}$, $p_a, p_{a+1}, \cdots, p_{n-2}, p_{n-1}$, and p_n are in clockwise order. If three or more endpoints are assessed relative to p_a, the method examines whether the connected endpoints exist inside area A'. Then, if the connected endpoints exist inside area A', the method examines whether they compose a convex polygon. If either condition is not satisfied, the method discards the endpoint that last assessed and the next endpoint is assessed. Let p_d be the first discarded endpoint by the latter examination. The method stores p_d for the next convex polygon to be enumerated. After assessing endpoints up to p_{a-1} (if $a = 1$, then up to p_n), the method calculates the size of the obtained convex polygon. This polygon is the first one of the polygons to be enumerated. Then, for the next convex polygon, the method connects p_a and p_{d-1} (if $d = 1$, then p_n). If $\overline{p_a p_{d-1}}$ exists inside A', then $p_{d'} = p_{d-1}$. Otherwise, the method looks at the continuing endpoint of p_{d-1} until it finds endpoint $p_{d'}$ where $\overline{p_a p_{d'}}$ exists inside A'. After finding $p_{d'}$, the method assesses the continuing endpoint of $p_{d'}$ and creates convex polygons while performing the abovementioned examinations. Moreover, storing the first discarded endpoint p_d by the latter examination and assessing from p_{d-1} and the continuing endpoints for the next convex polygon are performed in the manner described above. However, the new candidates p_{d_new} for p_d are limited to $d\_new > d$ or $1 < d\_new < a - 1$. The termination conditions for starting endpoint p_a are a new p_d not stored or if the number of endpoints from $p_{d'}$ to p_a is less than three. This process is performed repeatedly for each of all endpoints as the starting endpoint p_a.

After enumerating the number of convex polygons, the method finds a convex polygon c_m with maximum size. Moreover, the method finds a convex polygon c_l

with maximum size such that the center of the label position obtained using the proposed method exists inside the polygon. Note that polygons are excluded if the length of the long side of the minimum bounding rectangle exceeds five times that of the short side because such polygons are unsuitable for internal label placement.

3.3 Experimental Results

Let $dist$ be the distance between the centroid of c_l and the center of the label position obtained by the proposed method, and let $diag$ be the length of the diagonal of the minimum bounding rectangle of A'. The experimental results are shown in Table 1.

The value of c_l/c_m is high; thus, the proposed method can place a label inside a fairly large convex polygon. The value of $dist/diag$ is small; thus, the proposed method can place a label near the centroid of c_l. Note that the run time does not include the preprocessing steps. The proposed method is quite fast and can determine approximately tens of thousands of label positions in 100 ms, as requested by the employees of the Dawn Corporation.

We show label placement examples obtained by the proposed method in Figs. 5 and 6. Figure 5 shows the label placement for Aisai-shi (located in Aichi prefecture, Japan). The number of boundary lines of that was 2255. Here, the values of c_l/c_m and $dist/diag$ were 100 and 3.80%, respectively. As can be seen, the proposed method can find fairly good label positions for an area with some indents and narrow parts. Figure 6 shows the label placement for Neyagawa-shi (located in Osaka prefecture, Japan), which is not included in Table 1 because it has fewer

Table 1 Experimental results (c_l/c_m means how large convex polygon found by the proposed method; $dist/diag$ means how close the label position determined by the proposed method is to the centeroid of the convex polygon c_l)

Label	Number of boundary lines	c_l/c_m (%)	$dist/diag$ (%)	Run time (ms)
Aisai-shi	2255	100.0	3.80	0.0043
Ajigasawa-cho	6568	100.0	9.91	0.0042
Ashoro-cho	12,849	100.0	9.13	0.0075
Iida-shi	10,051	100.0	2.88	0.0039
Ishikari-shi	10,304	98.6	1.32	0.0046
Iyo-shi	4760	100.0	10.81	0.0042
Esashi-cho	4902	88.5	16.08	0.0039
Ozu-shi	6593	100.0	20.15	0.0042
Otsu-shi	4627	63.2	1.97	0.0044
Onojo-shi	1799	91.9	0.01	0.0043
average	6470.8	94.2	7.61	0.0046

Fig. 5 Label placement for
Aisai-shi

Fig. 6 Label placement for
Neyagawa-shi

indents and is suitable for label placement. The number of boundary lines of that
was 3586. As can be seen, the proposed method can find fairly good label positions
for an area with fewer indents and narrow parts.

4   Extending the Proposed Method

In this section, we consider how to place a label for an area when only a part of the
area is visible. In addition, we consider how to handle areas with inclusion
relations.

4.1   Label Placement for a Partial Area

By examining the following conditions, it is possible to determine whether A exists
inside the display frame, only a part of A is displayed, or A exists outside the display
frame.

(i)   Whether the boundary lines of A and the display frame intersect.
(ii) Whether an arbitrary point p within A exists inside the display frame.

Condition (i) can be determined quickly using the preprocessing steps described in Sect. 2.3. If any intersection exists, then only a part of the area is displayed. If no intersections exist and condition (ii) is satisfied (not satisfied), then A exists inside (outside) the display frame. Condition (ii) can be determined quickly by drawing a half line from p in one direction and counting the number of intersections with the display frame. If the number of intersections is an even value, p exists outside the display frame. If the number of intersection is an odd value, p exists inside the display frame.

Assume that only a part of area A exists inside display frame D, as shown in Fig. 7. When applying the proposed method to this case, we first find the intersections of the boundary lines of A and display frame D. Here we explain the bottom side of D as an example. As can be seen in Fig. 7, we treat the lower side of D as a horizontal auxiliary line h and find the intersections ch_1, ch_2, ... in a manner similar to that described in Sect. 2.2. In addition, we find intersections of h and the other sides of D, and we treat them in the same manner as the intersections of h and the boundary lines. Note that, in Fig. 7, the lower left and right corners are found. Then, we delete intersections that exist outside D or A. The white circles in Fig. 7 indicate that the intersections to be deleted. We then draw vertical line v' and create a label candidate in the same manner described in Sect. 2.2, as shown in Fig. 8. Note that with regard to v', the intersections include those of v' and the side of D, and intersections that exist outside D or A will be deleted.

Fig. 7 Horizontal *auxiliary line* for a partial area

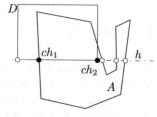

Fig. 8 Label candidate for a partial area

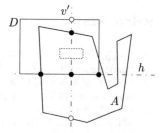

Fig. 9 *Auxiliary lines* for areas with inclusion relations

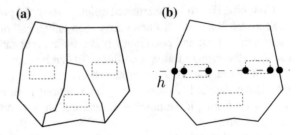

4.2   Label Placement for Areas with Inclusion Relations

Here we consider how to address areas with inclusion relations such as prefectures and cities. In such cases, if possible, overlapping labels for prefecture and city names should be avoided. When applying the proposed method to this case, we first place area labels for the lowest level of the inclusion relations, as shown in Fig. 9a. Then, we place labels for areas at the next level of the inclusion relations. At this time, we consider the placed labels as existing outside the areas, as shown in Fig. 9b, and the intersections of the auxiliary line and the placed labels are treated in the same manner as the intersections of the auxiliary and boundary lines. We continue this process recursively up to the highest-level areas. Thus, the proposed method can place labels for areas with inclusion relations while avoiding overlapping labels.

5   Conclusion

We have proposed a labeling method for areas on a map. The experimental results demonstrate that the proposed method is effective and can be executed in real-time. In future, we plan to develop a method to create more effective label candidates by employing the auxiliary lines that consider the shape of the area.

References

1. Hirsch, S.A.: An algorithm for automatic name placement around point data. Am. Cartographer **9**(1), 5–17 (1982)
2. Christensen, J., Marks, J., Shieber, S.: An empirical study of algorithms for point-feature label placement. ACM Trans. Graph. **14**(3), 203–232 (1995)
3. Kakoulis, K.G., Tollis, I.G.: A unified approach to labeling graphical features. In: Proceedings of the 14th Annual Symposium on Computational Geometry, pp. 347–356. (1998)
4. Wagner, F., Wolff, A., Kapoor, V., Strijk, T.: Three rules suffice for good label placement. Algorithmica **30**(2), 334–349 (2001)

5. Funakawa, K., Abe, N., Yamaguchi, K., Masuda, S.: Algorithms for the map labeling problem with priorities (in Japanese). IEICE Trans. Fund. Electron. Commun. Comput. Sci. J88-A(5), 677–681 (2005)
6. Kameda, T., Imai, K.: Map label placement for points and curves. IEICE Trans. Fund. Electron. Commun. Comput. Sci. E86-A(4), 835–840 (2003)
7. Abe, N., Kusaki, M., Masuda, S., Yamaguchi, K.: An algorithm for placing labels of the points and chains on a map. Int. J. Inf. Sci. Comput. Math 4(2), 79–99 (2011)
8. Rylov, M., Reimer, A.: A practical algorithm for the external annotation of area features. Cartographic J. 54(1), 61–76 (2017)
9. Aonuma, H., Imai, H., Tokuyama, T.: Some Voronoi diagrams for character placing problems in map databases (in Japanese). IPSJ SIG Tech. Rep., 89-AL-10-5, pp. 1–7. (1989)
10. Dörschlag, D., Petzold, I., Plümer, L.: Placing objects automatically in areas of maps. In: Proceedings of the 21st International Cartographic Conference, pp. 269275. (2003)
11. Fundamental Geospatial Data by the Geospatial Information Authority of Japan. http://www. gsi.go.jp/kiban/

Heuristic Test Case Generation Technique Using Extended Place/Transition Nets

Tomohiko Takagi, Akinori Akagi and Tetsuro Katayama

Abstract This paper shows a novel heuristic test case generation technique using an extended PN (place/transition net). A guard and action are introduced to improve the representation power of a PN. Also, a weight that represents the degree of testing priority is given to each transition of a PN. Our algorithm that heuristically searches the extended PN generates a specified number of feasible test cases that focus on parts of higher testing priority.

Keywords Model-based software testing · Place/transition net · Test case generation · Heuristic approach

1 Introduction

The recent spread of low-cost and high-performance hardware and networking facilities has stimulated software engineers to develop advanced concurrent software. The concurrent software includes multiple components that concurrently work and interact each other, which often results in an extremely complex behavior.

T. Takagi (✉)
Faculty of Engineering,
Kagawa University, 2217-20 Hayashi-Cho, Takamatsu-Shi, Kagawa 761-0396, Japan
e-mail: takagi@eng.kagawa-u.ac.jp

A. Akagi
Graduate School of Engineering, Kagawa University, 2217-20 Hayashi-Cho, Takamatsu-Shi, Kagawa 761-0396, Japan
e-mail: s17g451@stu.kagawa-u.ac.jp

T. Katayama
Institute of Education and Research for Engineering, University of Miyazaki, 1-1 Gakuen-Kibanadai-Nishi, Miyazaki-Shi, Miyazaki 889-2192, Japan
e-mail: kat@cs.miyazaki-u.ac.jp

© Springer International Publishing AG 2018
R. Lee (ed.), *Applied Computing & Information Technology*,
Studies in Computational Intelligence 727, DOI 10.1007/978-3-319-64051-8_7

Therefore, it needs to be systematically tested for the achievement of desired software reliability. MBT (model-based testing) is a systematic method to develop test cases on the basis of formal models that represent the expected behavior of SUT (software under test). Lots of researches on its area have been made [14], and the various kinds of techniques constructed by them have been tried to apply in actual software developments. The modeling language and notation in MBT is selected based on the characteristic of SUT and aspect to be tested, and Petri nets can be selected to test the behavior in the state space of concurrent software [12]. However, MBT using the Petri nets includes the following problems:

- The representation power of Petri nets is not enough to define the complex behavior of concurrent software.
- It is not easy to construct effective test cases for improving software reliability within a limited amount of time that has been allocated for test processes.

In order to address these problems, this paper shows a novel heuristic test case generation technique using an extended PN (place/transition net). The PN is a kind of Petri nets, and we extend it by introducing guards, actions, and weights.

The guard is a statement to represent additional conditions to enable the firing of a transition, and the action is a set of statements to represent the detailed procedure for data processing that needs to be performed just after a transition has fired. The guards and actions are useful to improve the representation power of a PN. In this study, they are written in VDM++ that is one of formal specification description languages for VDM (Vienna development method) [5].

Also, the weight is a value that is given to each transition of a PN, and it represents the degree of testing priority (for example, fault proneness, frequency of use, and so on). The value is calculated based on one or more kinds of software metrics. Finally, our algorithm that heuristically searches the extended PN generates effective test case, that is, a specified number of feasible test cases that focus on parts of higher testing priority.

The rest of this paper is organized as follows. Section 2 shows the definition and construction method of an extended PN, and then in Sect. 3 we propose a heuristic test case generation algorithm using the extended PN. Section 4 gives consideration based on an application example and related work. In the application example, an extended PN is constructed based on non-trivial software specifications, and it is used to generate test cases. Finally, Sect. 5 shows conclusion and future work.

2 Extended Place/Transition Nets

This section shows the definition and the construction method of an extended PN.

Fig. 1 Simple example of a PN (place/transition net)

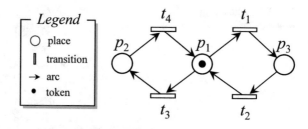

initial marking: [1, 0, 0]
final marking: randomly selected

2.1 Traditional PNs

Figure 1 is a simple example of a PN. A PN is a simple and fundamental formal model that can be utilized to define the expected behavior of SUT, and it is constructed based on specifications of the SUT. The PN consists of the following four kinds of elements:

- *places* to represent states of each component of SUT,
- *transitions* and *arcs* to represent state transition that is done by event occurrence in SUT, and
- *tokens* to indicate current states of components of SUT.

A state of overall SUT is represented as a *marking*, that is, distribution of tokens on a PN. The state transition of SUT is expressed as a 3-tuple list $<m_{from}, t, m_{to}>$, where m_{from} is a marking that means a from-state, t is a transition that fires in m_{from}, and m_{to} is a marking that means a to-state as a result of the firing of t. Note that the terms *transition* and *state transition* are distinguished in this paper; the former means a transition as an element of a PN, and the latter means state transition of overall SUT, that is, the above-mentioned 3-tuple list. An initial state of SUT is represented as an initial marking of a PN. For example, the initial marking of Fig. 1 is $[1, 0, 0]$, which means that the places p_1, p_2 and p_3 hold 1, 0 and 0 token, respectively. Also, in this study, final markings are defined to represent final states of SUT. Thus a test case based on a PN is created as a sequence of successive state transitions that starts with an initial marking and ends with a final marking. An example of a test case that can be created from Fig. 1 is $<[1, 0, 0], t_1, [0, 0, 1]> \rightarrow <[0, 0, 1], t_2, [1, 0, 0]>$.

2.2 Extended PNs

In order to improve the representation power of a PN, *guards*, *actions*, and *weights* are introduced into the PN in this study. The PN that includes the guards, actions,

Fig. 2 Simple example of an extended PN

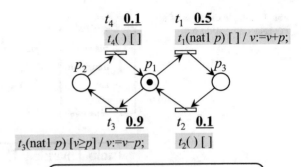

and weights is called an *extended PN*, and it is utilized to generate test cases in the algorithm proposed in next section. A simple example of an extended PN is shown in Fig. 2.

In a traditional PN, a transition can fire so long as all of its incoming places holds tokens, which is not enough to represent the complex behavior of practical SUT. A guard is additional conditions to enable the firing of a transition, and it is defined as a conditional statement within a square bracket that is included in a label for the transition in a PN. In this study, a guard is written in VDM++ that is one of formal specification description languages for VDM. VDM++ is useful to construct abstracted specifications that can be interpreted by computers, and therefore a PN and VDM++ are well suited with each other. For example, t_3 in Fig. 2 has a guard "$v \geq p$", which means that t_3 cannot fire when "$v \geq p$" is not satisfied. Note that p is an event parameter (an argument of t_3) of a *nat1* (natural number excluding 0) type that corresponds to input data from a user or another system to SUT, and v is an instance variable whose value can be changed in actions.

An action is a set of statements to formally represent the detailed procedure for data processing that needs to be performed just after a transition has fired. Event parameters of a corresponding transition can be read, and instance variables can be read, written, and newly defined in an action. A traditional PN provides no ways to represent such a detailed procedure. In this study, an action is defined at the tail of a label for a transition, and is written in VDM++. For example, t_3 in Fig. 2 has an action "$v := v - p;$".

A weight is a value that is given to each transition of a PN, and it represents the degree of testing priority (for example, fault proneness, frequency of use, and so on). The value is calculated based on one or more kinds of software metrics, and therefore, the meaning of the term *testing priority* is determined by selected software metrics. Test engineers need to select software metrics that

- are closely related to the testing priority in SUT,
- can be calculated from SUT and its development project, and
- can be exactly related to each transition of a PN.

Examples of software metrics that can be generally utilized to calculate the weights are LOC, cyclomatic complexity [3], the fault prediction using the past record of bug fixes [15], the fault prediction using identifiers in source codes [9], the fault prediction using majority ranking fuzzy clustering [1], and the probability distributions of operational profiles [10].

The procedure to calculate the weights is as follows:

(Step 1) Test engineers select software metrics, and also determine the degree of importance of each of the selected software metrics. They are performed based on the characteristics of SUT and its development project. A pairwise comparison method using questionnaires [2] can be applied to consider the importance.

(Step 2) The selected metrics are measured for each transition of a PN, and then are normalized between 0.0 and 1.0.

(Step 3) A weight for each transition of a PN is given by the following formula:

$$weight(t) = \sum_{i=1}^{N} (metrics_i(t) \times importance_i) \tag{1}$$

where t is a transition of a PN, N is the number of selected software metrics and satisfies $N \geq 1$, $metrics_i(t)$ represents the normalized value of the ith selected software metrics for t, and $importance_i$ expresses the degree of importance of the ith software metrics that satisfies the following constraints:

$$\begin{cases} \sum_{i=1}^{N} importance_i = 1.0 \\ 0.0 \leq importance_i \leq 1.0, \quad (i = 1, 2, \cdots, N) \end{cases} \tag{2}$$

In Fig. 2, the transitions t_1, t_2, t_3 and t_4 are given the weights 0.5, 0.1, 0.9 and 0.1 respectively.

3 Test Case Generation Algorithm

In this section, we propose the heuristic test case generation algorithm using the extended PN shown in the previous section.

This algorithm generates a specified number of feasible test cases in descending order of testing priority. A heuristic approach is introduced into this algorithm in order to generate effective test cases, that is, test cases that are feasible and have higher testing priority. The feasibility problem is caused by the dependence among guards and actions in an extended PN. For example, in Fig. 2, when the extended PN is in $[1,0,0]$ and $v=0$, a state transition $<[1,0,0],t_3,[0,1,0]>$ is not feasible, since there are no natural numbers excluding 0 (that is, values of a *natl* type) for the event parameter p that satisfy the guard "$v \geq p$". On the other hand, the testing priority is introduced to detect failures of SUT within a limited amount of time in actual software development processes, and it is calculated based on weights of an extended PN. The search space for test case generation in this study is extremely large, and therefore it can be achieved by heuristic approaches rather than deterministic approaches. Note that, in this algorithm, the feasibility of test cases is assured on an extended PN, but the priority of test cases is determined as approximate solutions.

The input of this algorithm are as follows:

- *PN*, that is, a traditional PN that represents the expected behavior of SUT: It does not include guards and actions.
- *EPN*, that is, an extended PN that represents the expected behavior of SUT: It includes *PN*.
- n_T, that is, the number of test cases to be generated: It should satisfy $n_T \geq 1$, and is determined based on the effort that can be devoted to this technique.
- n_C, that is, the number of test case candidates from which this algorithm selects the best one: It should satisfy $n_C \geq 2$.
- r, that is, the number of consecutive times to permit to retry the generation of test case candidates when the generation fails: It should satisfy $r \geq 0$.
- N, that is, the length of successive state transitions to be covered by test cases: It is based on the idea of N-switch testing techniques [12].

n_C and r are used to adjust the cost and performance of this algorithm. The larger n_C and r are, the more this algorithm can generate effective test cases and also uses computer resources.

The output of the algorithm is T, that is, a list of test cases that consists of n_T and less test cases. T is empty when this algorithm starts.

This algorithm consists of the following eight steps:

(Step 1) A current marking (a marking that is currently considered in this algorithm) is set to an initial marking of *EPN*. For example, when Fig. 2 is inputted as *EPN*, a current marking is set to $[1,0,0]$. Also, instance variables are set to their initial values, if necessary.

(Step 2) A set of transitions that are fireable in the current marking are derived from the *PN*. For example, when the current marking is $[1,0,0]$ in Fig. 2 (interpreted as *PN*), the set is $\{t_1,t_3\}$.

(Step 3) If the set of fireable transitions is empty, this algorithm terminates since *EPN* would include errors. If it is not empty, a transition is randomly selected according to weights from the set of fireable transitions, and then the feasibility of the selected transition is checked on *EPN*. If the selected transition has event parameters, their values are generated based on predefined manners.

For example, when the set is $\{t_1, t_3\}$ in Fig. 2, either of the transitions is selected by about 36% (5/14) and 64% (9/14) respectively, and its feasibility is checked on Fig. 2.

(Step 4) If the feasibility of the selected transition has been confirmed, the firing of the selected transition is determined (that is, the current marking is changed, and the actions of the selected transition are executed). If the feasibility has not been confirmed, the selected transition is removed from the set of fireable transitions, and this algorithm returns to Step 3.

For example, in Fig. 2, when the current marking, the set of fireable transitions, and the instance variable are $[1, 0, 0]$, $\{t_1, t_3\}$, and $v = 0$ respectively, t_1 is feasible but t_3 is not feasible. Therefore, if t_1 has been selected in Step 3, the current marking is changed to $[0, 0, 1]$, and its action "$v := v + p$;" is executed (if the event parameter p is 10, the value of v is changed from 0 to 10). On the other hand, if t_3 has been selected in Step 3, the set of fireable transitions is changed from $\{t_1, t_3\}$ to $\{t_1\}$, and then this algorithm returns to Step 3.

(Step 5) If the current marking is not a final marking, this algorithm returns to Step 2.

(Step 6) The obtained feasible successive state transition sequence that starts with an initial marking and ends with a final marking on *EPN* is added to C. C is a set of test case candidates, and it is empty when this algorithm starts. If the number of elements of C has not reached n_C, this algorithm returns to Step 1.

(Step 7) Each element of C is evaluated by the following formula:

$$evaluation(t_C) = \frac{sum(t_C)}{length(t_C)} \tag{3}$$

where $t_C \in C$, $length(t_C)$ expresses the number of transitions that appear in t_C, and $sum(t_C)$ expresses the sum of weights of transitions that satisfy the following (a) or (b):

(a) The transition is the tail of a successive state transition sequence of length N that is included in t_C, and the sequence has not been covered by the elements of T.

(b) The transition is the tail of a successive state transition sequence of length N and less. The sequence starts with the initial marking, and is included in t_C. Also, the sequence has not been covered by the elements of T.

The successive state transition sequence discussed in (a) and (b) is called a *measuring object*. Traditional N-switch testing techniques [12] are intended to cover all of the feasible measuring objects, but our technique is not necessarily intended to it.

For example, in Fig. 2, when $t_C = \ <[1, 0, 0], t_1, [0, 0, 1]> \ \rightarrow \ <[0, 0, 1]$, $t_2, [1, 0, 0]> \ \rightarrow \ <[1, 0, 0], t_1, [0, 0, 1]> \ \rightarrow \ <[0, 0, 1], t_2, [1, 0, 0]>$,

$T = \{ <[1,0,0], t_1, [0,0,1] > \rightarrow <[0,0,1], t_2, [1,0,0] > \}$ and $N = 2$, t_C is evaluated by $evaluation(t_C) = 0.5/4 = 0.125$. The numerator 0.5 corresponds to the weight of t_1 (the tail of a successive state transition $<[0,0,1], t_2, [1,0,0] > \rightarrow <[1,0,0], t_1, [0,0,1] >)$.

An element of the highest evaluation in C is added to the tail of T (if the highest evaluation is not 0), and then C is initialized to an empty set.

(Step 8)   If the number of elements of T does not reach n_T, and the number of consecutive times that the highest evaluation in C results in 0 (that is, effective test case candidates fail to be generated) does not reach $r + 1$, then this algorithm returns to Step 1.

4 Discussion

This section gives consideration based on an application example and related work. In the application example, an extended PN is constructed based on non-trivial software specifications, and it is used to generate test cases.

4.1 Application Example

In this study, we have constructed an extended PN that represents the expected behavior of an EMCS (electronic money charging system). Its overview is shown in Fig. 3. Note that the extended PN of the EMCS is large, and therefore some parts

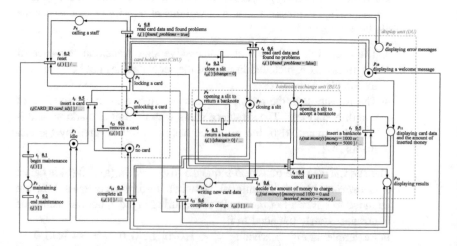

Fig. 3 Overview of an extended PN of an EMCS (electronic money charging system)

are omitted to save the space in this paper. The EMCS can accept an IC card and banknotes, and then register a specified amount of money to the IC card. The EMCS includes data processing such as the calculation of the amount of money to be charged and returned. It needs to be defined as actions and guards, and can hardly be represented in a traditional PN.

Also, we have developed a prototype of a test case generation tool based on this technique.

In order to evaluate this technique, we generated test cases by applying the tool to the extended PN of the EMCS with $n_T = 50$, $n_C = 30$, $r = 10$, and $N = \{1, 2, 3, 4, 5\}$. Figure 4 shows the growth of cumulative evaluation of generated test cases, and Fig. 5 shows the evaluation of each test case that is calculated by the formula (3). A larger value of N results in a larger number of test cases. In this application example, the test case generation is completed before the number of generated test cases reaches n_T, since the number of consecutive times that effective test case candidates fail to be generated reaches $r + 1$ in Step 8 (see the previous section). It is found that the cumulative evaluation grows rapidly on earlier stage of the generation, that is, test cases of higher evaluation (higher testing priority) are

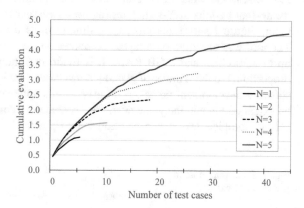

Fig. 4 Growth of cumulative evaluation of generated test cases

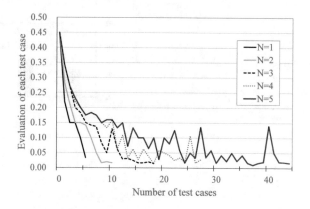

Fig. 5 Evaluation of each test case

Fig. 6 Growth of the
cumulative number of found
measuring objects

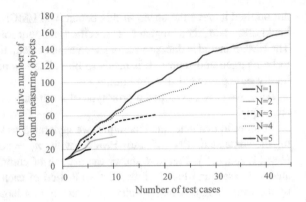

preferentially generated. Therefore test engineers will perform effective testing,
even if they cannot perform all of the generated test cases due to the limitation of
time that has been allocated for test processes. Figure 6 shows the growth of the
cumulative number of found measuring objects. Its curves are similar to those of
Fig. 4. A larger value of N results in a larger number of measuring objects, which
means that traditional N-switch testing needs a larger number of test cases to cover
them. Figure 7 shows the processing time (sec) of the tool to generate test cases.
The tool was executed on a laptop computer with i7-4650U processor (1.70 GHz,
up to 3.30 GHz) and 8 GB RAM. It is found that the processing time is roughly
proportional to the number of generated test cases, and it will be allowable in actual
software development processes, at least in this application example. If the size of
an extended PN and the value of n_T, n_C, r, N are extremely large, a tool will not be
able to complete test case generation within practical time.

Fig. 7 Processing time to
generate test cases

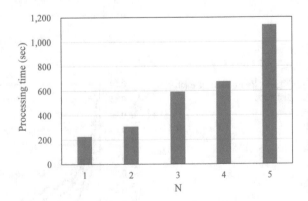

4.2 Related Work

There are some previous studies that relate to MBT using Petri nets. For example, Li et al. [8] show a systematic interoperability testing technique of MIPv6 based on Petri nets. In the technique shown by Dong et al. [4], high-level Petri nets are utilized to test BPEL (business process execution language for Web services)-based Web service composition. Also, a literature [12] shows the technique to generate test cases based on a finite state machine generated from a PN; it is intended to cover all the successive state transition sequences of specified length, and therefore a large number of test cases often need to be executed to satisfy a selected test stopping criterion. The technique proposed in this paper is intended to concentrate on state transitions of higher testing priority, rather than to cover all the state transitions. As regards the feasibility problem, it is one of the most important themes in MBT. Recent studies report that heuristic approaches can be effectively applied to solve it and to optimize test cases. For example, Kalaji et al. [7] show a technique to generate feasible test cases from an extended finite state machine by the application of a genetic algorithm. We proposed a technique to generate test cases from an extended operational profile by using ant colony optimization [11], but finite state machines are not necessarily suitable for modeling the behavior of concurrent software.

The difficulty of this technique is how to relate software metrics to each transition of a PN. This technique requires establishing the traceability between a PN and materials that are used for calculating software metrics, especially source codes of SUT. One of the ultimate ways to solve it is to introduce round-trip engineering [6], that is, to construct an environment that enables the source code generation from a PN (forward engineering) and the PN generation from the source codes (reverse engineering). If the direct relationships between a PN and materials are kept by the round-trip engineering, each transition is automatically given the values of software metrics that are measured by tools.

This technique can be a foundation for model-based mutation testing [13]. For example, mutation operators can be applied to parts of higher testing priority in order to generate effective mutant PNs that include realistic latent failures. The mutant PNs are utilized to evaluate the quality of test cases (mutation analysis) and to generate negative test cases that focus on specific faults that are caused by such failures (negative testing).

5 Conclusion and Future Work

This paper has shown a novel heuristic test case generation technique using an extended PN. A guard and action are introduced to improve the representation power of a PN. Also, a weight that represents the degree of testing priority is given to each transition of a PN. Our algorithm that heuristically searches the extended

PN generates a specified number of feasible test cases that focus on parts of higher testing priority. An application example suggests that test engineers will perform effective testing, even if they cannot perform all of the generated test cases due to the limitation of time that has been allocated for test processes. The processing time to generate test cases by a tool will be allowable in actual software development processes, if the size of an extended PN and the values of input parameters for the algorithm are not large. However, the construction of an extended PN will be a heavy task for some test engineers.

Our future work includes the enhancement of performance of the test case generation algorithm, the development of a technique to support the construction of extended PNs, and so on. Additionally, we plan to apply the improved technique to large-scale software and evaluate its effectiveness again.

Acknowledgements This work was supported by JSPS KAKENHI Grant Number 26730038.

References

1. Abaei, G., Selamat, A.: Increasing the accuracy of software fault prediction, using majority ranking fuzzy clustering. Int. J. Softw. Innov. **2**(4), 60–71 (2014)
2. Arao, T., Machida, Y., Toda, K., Yaegashi, R., Takagi, T.: Decision-making, about software release time using analytic hierarchy process. In: Proceedings, of 3rd International Conference on Advanced Applied Informatics, pp. 751–756 (2014)
3. Beizer, B.: Software Testing Techniques, 2nd edn. Van Nostrand Reinhold (1990)
4. Dong, W.L., YU, H., Zhang, Y.B.: Testing BPEL-based Web service composition using high-level Petri nets. In: Proceedings of 10th International Enterprise Distributed Object Computing Conference, pp. 441–444 (2006)
5. Fitzgerald, J., Larsen, P.G., Mukherjee, P., Plat, N., Verhoef, M.: Validated Designs for Object-Oriented Systems. Springer-Verlag, London (2005)
6. Frankel, D.S.: Model Driven Architecture: Applying MDA to Enterprise Computing. John Wiley & Sons (2003)
7. Kalaji, A., Hierons, R.M., Swift, S.: Generating feasible transition paths for testing from an extended finite state machine (EFSM). In: Proceedings of International Conference on Software Testing Verification and Validation, pp. 230–239 (2009)
8. Li, H., Ye, X., Wu, C., Liu, L., Wang, L.: Modeling interactive property of MIPv6 with Petri net for interoperability testing. In: Proceedings of 2nd International Conference on Information and Computing Science, pp. 313–316 (2009)
9. Mizuno, O., Kawashima, N., Kawamoto, K.: Fault-prone module prediction approaches using identifiers in source code. Int. J. Softw. Innov. **3**(1), 36–49 (2015)
10. Musa, J.D.: The operational profile. In: Reliability and Maintenance of Complex Systems, *NATO ASI Series F: Computer and Systems Sciences*, vol. 154, pp. 333–344 (1996)
11. Takagi, T., Beyazıt, M.: Optimized test case generation based on operational profiles with fault-proneness information. Stud. Comput. Intell. **578**, 15–25 (2015)
12. Takagi, T., Oyaizu, N., Furukawa, Z.: Concurrent N-switch coverage criterion for generating test cases from place/transition nets. In: Proceedings of 9th International Conference on Computer and Information Science, pp. 782–787 (2010)

13. Takagi, T., Takata, R., Furukawa, Z., Belli, F., Beyazıt, M.: Metrics for modelbased mutation testing based on place/transition nets. In: Proceedings of Joint Conference of 21st International Workshop on Software Measurement and 6th International Conference on Software Process and Product Measurement, pp. 7–10 (2011)
14. Utting, M., Pretschner, A., Legeard, B.: A taxonomy of model-based testing approaches. Softw. Test. Verifi. Reliab. **22**, 297–312 (2012)
15. Whittaker, J.A., Arbon, J., Carollo, J.: How Google Tests Software. Addison-Wesley Professional (2012)

Risk Assessment of Security Requirements of Banking Information Systems Based on Attack Patterns

Krissada Rongrat and Twittie Senivongse

Abstract Security risk assessment is an important process for the implementation of any information systems including those in the banking sector. When a bank initiates or implements an information system project, requirements engineers or business analysts in the project conduct an initial validation of system security requirements to check if they comply with banking security regulations before an audit takes place. This paper presents an initial risk assessment method to assist the project team in validating security requirements of a banking information system. Text similarity analysis is used to identify which security regulations are missing from the security requirements of the bank, and a quantitative risk index model is also proposed to determine the level of risk associated with the regulations missing from the requirements. The risk level is based on the harm any potential attacks can do to the information system if the missing regulations are not implemented. Using a case study of banking in Thailand, we apply the method to assess security requirements of Thai commercial banks against the IT Best Practices of the Bank of Thailand. We evaluate the performance of security compliance checking in terms of F-measure and accuracy, and validity of risk assessment in terms of correlation with security expert judgment.

Keywords Security requirement · Risk assessment · Attack pattern · Regulatory compliance · Text similarity · Banking

K. Rongrat · T. Senivongse (✉)
Department of Computer Engineering, Faculty of Engineering, Chulalongkorn University, Bangkok 10330, Thailand
e-mail: twittie.s@chula.ac.th

K. Rongrat
e-mail: krissada.ro@student.chula.ac.th

© Springer International Publishing AG 2018
R. Lee (ed.), *Applied Computing & Information Technology*,
Studies in Computational Intelligence 727, DOI 10.1007/978-3-319-64051-8_8

1 Introduction

Security is one of the most crucial attributes that must be taken into account during the development of any information system. In the banking sector, there is an increase in the number, sophistication, and scope of cyber attacks against the industry [1]. It is essential that security concern needs to be part of the development of any banking information system from the beginning. Therefore, security requirements of the system must address security matters in a complete controlled structured way on the basis of recognized standards and best practices.

This paper uses the case of the banking sector in Thailand as a case study. Security requirements of an information system of a commercial bank in Thailand have to comply with a set of regulations called IT Best Practices [2, 3]. The IT Best Practices is a regulatory agreement between the Bank of Thailand (BOT) and commercial banks in Thailand to ensure that when any commercial bank needs to develop or customize an information system, security requirements of the system must be validated to check their compliance with the IT Best Practices early at the beginning of the project before proceeding to development. On the other hand, the bank can develop the system first, but before launching it to production, the system security requirements must be validated. In normal practice, before the validation by internal auditors of the bank and auditors from the BOT, security requirements are validated initially by either the requirements engineers or business analysts of the project. Since the validation requires a study of textual security requirements and IT Best Practices to determine if the regulations are met, this consumes project time and cost and largely relies on knowledge and experience of the requirements engineers and business analysts. Misjudgment could mean that some regulations are merely partially met or even missing and it would be more costly to find that out later in the project or leave it until the auditors find out.

This paper presents a method to help requirements engineers and business analysts of a banking system project to assess the security requirements of the system to be developed. The assessment comprises (1) checking compliance with the IT Best Practices and (2) assessment of risk associated with non-compliant requirements. On checking compliance, text similarity analysis is used to determine which security practices are missing from the security requirements document. Given those missing practices, we determine potential security attacks that could occur and assess the degree of risk based on the harm those attacks can do to the system. To assess the risk, we use the CAPEC attack pattern classification [4] to build a risk index model. The assessment result identifies non-compliant locations within the security requirements document and the degree of risk of potential attacks if the system is implemented based on such incomplete requirements. We evaluate the performance of compliance checking as well as validity of risk assessment.

Section 2 of this paper presents important background of the work. Section 3 discusses related research. Section 4 describes the proposed risk assessment method, followed by a support tool in Sect. 5. An evaluation is shown in Sect. 6 and the paper concludes in Sect. 7.

2 Background

2.1 IT Best Practices

IT Best Practices [2, 3] is a document developed by the (BOT) and commercial banks in Thailand as a recommendation of security solutions to control risk that could occur to banking information systems. The recommendation is based on cybersecurity frameworks developed by NIST, ISO 27005, COBIT etc. and covers operation procedures, operation controls, and information systems. The risk control part of the IT Best Practices specifies baseline requirements for banking information systems and is the most relevant in the context of this paper. The baseline requirements address five domains of information systems: (1) Core Banking Application, (2) ATM Application Control, (3) ATM Machine, (4) Internet Banking Application Control, and (5) Internet Banking Security.

2.2 CAPEC Attack Patterns

Attack patterns document reusable attack knowledge to bridge the knowledge gap and assist with attack analysis [5]. The Common Attack Pattern Enumeration and Classification (CAPEC) [4] is a taxonomy of cyber security attacks developed by MITRE corporation. It has been incrementally built, starting from 2007, and includes 609 attack patterns thus far. Each attack pattern captures knowledge about how specific parts of an attack are designed and executed and gives guidance on how to mitigate the effectiveness of the attack. Each attack pattern description includes several topics, e.g. Summary, Attack Execution Flow, Typical Severity, Typical Likelihood of Exploit, Methods of Attack, Examples-Instances, Attackers Skills and Knowledge Required, Resources Required, Solutions and Mitigations, Related Weaknesses, Relevant Security Requirements, Confidentiality/Integrity/ Availability Impact, Technical Context. Among these, we use Solutions and Mitigations, Relevant Security Requirements, Typical Severity, and Typical Likelihood of Exploit information for risk assessment.

3 Related Work

In this section, we discuss related work on application of text analysis to software requirements and security risk assessment.

On application of text analysis to software specification, Stierna and Rowe [6] argue that finding opportunities for reuse of previously written software modules in large and complex systems is difficult. Instead, the reuse opportunities can be found indirectly through software requirements. That is, they match written requirements of the new software against the requirements used to define the old software, and requirement pairs with common words suggest reuse of software modules related to the old requirements. We follow their approach in processing textual software requirements and using Cosine coefficient [7] to measure the degree of similarity between requirements. Ilyas and Kung [8] present a requirement similarity measurement framework to support similarity measurement for the requirements of a running project and the requirements of the already completed projects. They use Dice, Jaccard, and Cosine coefficients as similarity measures. Once similar requirements are found, the design and code of the already completed projects become reusable components. Dag et al. [9] use an automated analysis of flow of software requirements to increase efficiency of their requirements engineering process. When there are new requirements coming continuously from many different sources and having to be responded quickly for short time-to-market, they need to identify relationships between requirements. They use text similarity analysis to find duplicate requirements so that they can avoid doing the same job twice, assigning the same requirement to different developers, or getting two solutions to the same problem. Also, they use Dice, Jaccard, and Cosine coefficients as similarity measures but Dice and Cosine coefficients perform better in the experiment.

On security risk assessment, Yu et al. [10] presents an automated tool to support the use of formal logic, i.e. security argumentation, to determine security satisfaction of security requirements, or arguments. The tool includes a Lucene-based search engine for security attacks and weaknesses information which is taken from CAPEC and CWE catalogs. Keywords from the arguments are searched for relevant attacks, weaknesses, and mitigations on which the assessment of risk level (i.e. likelihood x impact) is based. When the risks from the arguments are acceptable, the system is considered to have reached satisfactory security. As with this work, our assessment of risk of security requirements is based on information about potential attacks from CAPEC. Unlike this work, the assessment is based on non-compliance with the regulations. Other security assessment researches based on security catalogs are found also in different contexts, e.g. Piromsopa et al. [11] use web server vulnerability (or CVE) information from MITRE Corporation, issue HTTP requests to scan web servers to find vulnerabilities, and assess risk based on the probability and impact of each vulnerability on web servers. Banklongsi and Senivongse [12] use information from CAPEC to define a security metric for web services based on the percentage of countermeasures provided against a certain attack type as well as severity, likelihood of exploit, and impact of the attack type.

4 Security Risk Assessment Method

The overview of the security risk assessment method is depicted in Fig. 1. We compile a standard set of security requirements from the IT Best Practices and match them with CAPEC attack patterns via the solutions and relevant security

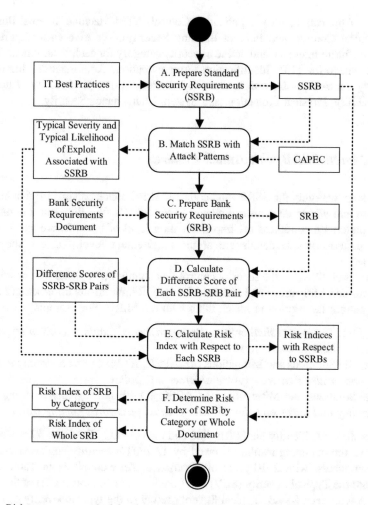

Fig. 1 Risk assessment method for the security requirements of a bank

requirements of the patterns. The matching helps identify severity and likelihood of exploit of the attacks that might occur if the standard requirements are missing from the bank security requirements. We then calculate risk indices for the bank security requirements. The details are as follows.

4.1 Prepare Standard Security Requirements (SSRB)

First, we extract standard security requirements of banking (SSRB) from the IT Best Practices [2, 3]. There are 52 standard requirements under five domains (i.e. Core

Banking Application, ATM Application Control, ATM Machine, Internet Banking Application Control, and Internet Banking Security). We give each requirement an ID for future reference and define a security category for each requirement. There are 12 categories [13]: Identification, Authentication, Authorization, Immunity, Integrity, Intrusion Detection, Non-repudiation, Privacy, Security Auditing, Survivability, Physical Protection, and System Maintenance Security.

4.2 Match SSRB with Attack Patterns

In this step, we study the SSRBs and collect CAPEC attack patterns [4] that involve software and whose contents are related to the SSRBs. We filter out a number of attack patterns that address the implementation level of the software and keep 24 attack patterns whose details are at the requirement level. Table 1 shows an example.

Since each SSRB is the security control that should be in place, we consider the Solutions and Mitigations and Relevant Security Requirements information in each attack pattern description to match them with the SSRB. For example,

SSRB_002: *The system shall enforce strong password (contain a mix of alphabetic and non-alphabetic characters).*
matches the Solutions and Mitigations of CAPEC_ID 16: *Create a strong password policy and ensure that your system enforces this policy.*
and the Solutions and Mitigations of CAPEC_ID 49: *Put together a strong password policy and make sure that all user created passwords comply with it.*

Therefore, we identify the SSRB_002 and the CAPEC_ID 16 and 49 as a match. We have the matching results reviewed by 12 BOT's security engineers and network engineers, with 2–10 years of experience. An example is in Table 2. We associate the Typical Severity (SEV) and Likelihood of Exploit (LOE) of the attack patterns with each SSRB. Typical Severity refers to the typical severity of impact on the software if this attack occurs. Typical Likelihood of Exploit means the likelihood of this attack typically succeeding considering the weakness attack

Table 1 Example of selected attack patterns

CAPEC_ID	Attack pattern name	SEV	LOE
16	Dictionary-based Password Attack	High	Medium
49	Password Brute Forcing	High	Medium
50	Password Recovery Exploitation	High	Medium
60	Reusing Session IDs (aka Session Replay)	High	High
94	Man in the Middle Attack	Very high	very high
169	Footprinting	Very low	High
...	...	...	...

Table 2 Example of SSRBs and associated attack patterns

SSRB_ID	SSRB description	CAPEC_ID	SEV	LOE
SSRB_001	The system shall use strong encryption and security protocols to safeguard sensitive data during transmission over open, public networks	94	5	5
SSRB_002	The system shall enforce strong password (contain a mix of alphabetic and non-alphabetic characters)	16, 49	4	3
SSRB_003	The system shall use two-factor authentication before password reset	60	4	4
SSRB_004	The system shall re-authenticate when customer performs change of profile (e.g. address, telephone number, email) by hardware token	169, 50	4	4
...	...	...	...	...

surface, skills and resources required, available blocking solutions etc. We map the ordinal scale of (Very Low, Low, Medium, High, Very High) of SEV and LOE to a numerical scale of 1–5 respectively for later calculation of the risk index. In the case that the SSRB is matched with more than one attack pattern, we use the principle of High Water Mark, i.e. the maximum level, to determine the SEV and LOE associated with the SSBR. For example, given Tables 1 and 2, the associated SEV of SSRB_004 is max(Very Low, High) = High = 4 and LOE is max(High, Medium) = High = 4.

4.3 Prepare Bank Security Requirements (SRB)

Given a security requirements document of a commercial bank, we prepare a list of bank security requirements (SRB), give each requirement an ID for future reference, and organize them into the same five domains as in the case of the SSRBs. An example is in Table 3.

Table 3 Example of SRBs

SRB_ID	SRB description
SRB_001	The application shall use AES encryption and SSL protocol to safeguard during transmission over public networks
SRB_002	The system will enforce a password to contain a mix of alphabetic and non-alphabetic characters and minimum password length is 8 characters
SRB_003	The application shall use two-factor authentication (OTP) before password reset
...	...

4.4 Calculate Difference Score of Each SSRB-SRB Pair

To determine how compliant the SRBs are with the SSRBs, we calculate the difference score of each SSRB and SRB requirement pair based on a text similarity measure. We preprocess the SSRB and SRB texts before similarity comparison as follows.

1. *Segment words and remove stop words*: For SSRBs and SRBs, perform word segmentation on each requirement and remove stop words taken from a list of 619 words from http://countwordsfree.com/stopwords, plus 39 common domain words, e.g. system, application, customer.
2. *Change to lowercase letters*: Change all capitalized words to lowercase except for words with specific meanings, e.g. SSL, HTTP, SNA.
3. *Remove suffixes*: Remove suffixes of words by truncating their suffixes using the Porter stemming algorithm.
4. *Determine difference between each requirement pair based on degree of similarity*: Using the vector space model, we represent each SSRB and SRB requirement as a weighted term vector, where each element w_i of the vector is the weight of the term (or word) i that appears in that requirement. We follow the Cosine coefficient to determine similarity between each SSRB-SRB requirement pair by using the Term Frequency-Inverse Document Frequency weight (TF-IDF) [7]. Since the similarity measure is bounded in [0, 1], it can be adapted to calculate the degree of difference $D_{q,r}$ between an SSRB requirement q from the IT Best Practices and an SRB requirement r of a bank by

$$D_{q,r} = 1 - \frac{\sum_{i=1}^{N} (w_{q,i} * w_{r,i})}{\sqrt{\sum_{i=1}^{N} w_{q,i}^2} * \sqrt{\sum_{i=1}^{N} w_{r,i}^2}} \tag{1}$$

where

$w_{q,i}$ weight of word i in SSRB requirement q,
$w_{r,i}$ weight of word i in SRB requirement r,
N number of distinct words in q and r, and
$D_{q,r}$ is in [0, 1].

Note that the following weight $w_{s,i}$ is used to determine the weight of word i in document s (i.e. each requirement q or r):

$$w_{s,i} = \begin{cases} tf_{s,i} * idf_i = (1 + \log_2 f_{s,i}) * \log_2 \frac{D}{d_i} & \text{if } f_{s,i} > 0 \\ 0 & \text{otherwise} \end{cases} \tag{2}$$

Table 4 Example of difference scores of SSRB-SRB pairs

SSRB_ID	SRB_ID	$D_{q,r}$
SSRB_001	SRB_001	0.3264
	SRB_002	1.0
	SRB_003	1.0
SSRB_002	SRB_001	1.0
	SRB_002	0.4116
	SRB_003	1.0
SSRB_003	SRB_001	1.0
	SRB_002	1.0
	SRB_003	0.0871
SSRB_004	SRB_001	1.0
	SRB_002	1.0
	SRB_003	1.0

where

D number of SRBs and
d_i number of SRBs in which word i appears.

These $D_{q,r}$ values are calculated for every pair of SSRB and SRB requirements. Given the SSRBs in Table 2 and SRBs in Table 3, the difference scores are shown in Table 4.

4.5   Calculate Risk Index with Respect to Each SSRB

A risk index is a product of the probability of a risk (i.e. likelihood) and the severity of impact caused by the risk (i.e. consequence). In our context, there is the probability of risk of attacks associated with an SSRB requirement when it is not met by any SRBs of a bank. We represent this probability of risk by the minimum $D_{q,ri}$ of each SSRB requirement q, i.e. the difference of the SRB requirement r_i that is the best match with the SSRB q. This probability is also weighted by the LOE of the attacks that could occur in the absence of that SSRB q from the banking information system. We use the SEV of those attacks as the severity impact caused by the risk.

Thus, a risk index R_q with respect to any SSRB requirement q is calculated by

$$R_q = min(D_{q,r}) \times LOE_q \times SEV_q \tag{3}$$

where

$min(D_{q,ri})$ minimum difference score of an SSRB requirement q and any SRB requirement r_i,
LOE_q Typical Likelihood of Exploit of attack patterns associated with q (e.g. LOE in Table 2),

Table 5 Example of risk indices when threshold is 0.25

SSRB_ID	SRB_ID	min(D_q,r_i)	SEV_q	LOE_q	R_q
SSRB_001	SRB_001	0.3264	5	5	8.16
SSRB_002	SRB_002	0.4116	4	3	4.9392
SSRB_003	SRB_003	0.0871 $\geq$ 0	4	4	0
SSRB_004	N/A	1.0	4	4	16

SEV_q Typical Severity of attack patterns associated with q (e.g. *SEV* in Table 2), and

R_q is in [0, 25].

For example, given Tables 2 and 4, the risk index with respect to each SSRB is in Table 5. We define a threshold (0.25 in this example) such that if the minimum difference score is greater than the threshold, we consider the SSRB *missing* from, i.e. not met by, the SRB requirements. Therefore, there is a degree of risk associated with the missing SSRB. On the other hand, if the minimum difference score falls below the threshold, we reset it to 0 and consider the SSRB *not missing* as an SRB that can meet that SSRB requirement is found.

4.6 Determine Risk Index of SRB by Category or Whole Document

In the previous section, an SRB that is the best match with an SSRB is identified, and a risk index is calculated with respect to how well the SRB can meet the SSRB. We then can determine the overall risk index for each of the 12 categories of the SRBs based on the risk indices associated with all SRBs under that category. Likewise, the overall risk index of the whole SRB document can be determined by the risk indices associated with all SRBs of a bank. Using the High Water Mark, we can represent the risk index R_c, where c is either a security category context or the whole document context, by the maximum risk index R_q associated with that context as in

$$R_c = max(R_q) \tag{4}$$

Given Table 5, the risk index by each category of the SRBs is shown in Table 6.

Table 6 Example of risk index by security category

SSRB_ID	SRB_ID	Security Category	R_q	R_c
SSRB_001	SRB_001	Integrity	8.16	8.16
SSRB_002	SRB_002	Authentication	4.9392	4.9392
SSRB_003	SRB_003	Authentication	0	
SSRB_004	N/A	Identification	16	16

```
<?xml version="1.0" encoding="UTF-8"?>
 <banksecurityreq>
  <ib>
   <requirement>
    <id>SSRB_001</id>
    <type>Integrity</type>
    <desc>The system shall use strong encryption and security protocols to
safeguard  sensitive  data  during  transmission  over  open,  public
networks.</desc>
    <serv>5</serv>
    <loe>5</loe>
   </requirement> ...
```

Fig. 2 SSRB in XML

5 Supporting Tool

A prototype tool is developed in Java to support the application of the proposed
security risk assessment. It supports the risk assessment model and automatically
compute risk indices related to an SRB document. The SSRB requirements are
represented in XML as in Fig. 2 and an SRB document of a bank is in docx or doc
format. The tool parses the files and processes to measure requirement differences
and calculate risk indices.

6 Evaluation

In this section, we report on the performance of compliance checking and on
validity of the risk index.

6.1 Performance of Compliance Checking

We use the security requirements documents of five commercial banks in Thailand
to evaluate how well the tool can identify which SSRBs are missing from the SRBs

of the banks. The tool calculates the difference score of each SSRB-SRB requirement pair, finds the SRB that is the best match, and determines if there are SSRBs
that are not met by any SRBs with regard to a threshold. We use the results of bank
requirements validation from the audits as the solution against which the performance of the tool is evaluated. The solution tells which SSRBs are considered by
the auditors as not met by, or missing from, the SRB documents of the banks.

Tables 7, 8, and 9 show the performance of the tool in terms of F-measure of the
two predicted classes of the SSRBs (i.e. Missing and Not Missing) and the overall
accuracy respectively. The result is depicted graphically in Fig. 3. At the threshold
of 0.4, all average performance reaches 80% and the best performance is when the
threshold is 0.5. Starting at the threshold of 0.52, false negatives begin to show (i.e.
the tool predicts missing SSRBs as Not Missing) and the F-measure of the Missing
class subsequently drops. We consider the false negatives as risky and not desirable,

Table 7 F-measure of Missing class

Threshold	Doc#1	Doc#2	Doc#3	Doc#4	Doc#5
0.05	38.10	88.24	11.11	41.67	87.50
0.15	57.14	93.75	14.29	50.00	90.32
0.25	72.73	96.77	25.00	58.82	93.33
0.30	88.89	96.77	28.57	62.50	93.33
0.40	88.89	93.33	50.00	71.43	96.55
0.50	88.89	92.86	100.00	76.92	96.55
0.55	75.00	88.89	N/A	90.91	96.55
0.65	85.71	80.00	N/A	90.91	92.86
0.75	85.71	50.00	N/A	88.89	88.00
0.85	40.00	N/A	N/A	57.14	72.73
0.95	40.00	N/A	N/A	33.33	44.44

Table 8 F-measure of Not Missing class

Threshold	Doc#1	Doc#2	Doc#3	Doc#4	Doc#5
0.05	31.58	33.33	27.27	12.50	50.00
0.15	76.92	75.00	53.85	50.00	66.67
0.25	89.66	88.89	81.25	69.57	80.00
0.30	96.77	88.89	84.85	75.00	80.00
0.40	96.77	80.00	94.44	84.62	90.91
0.50	96.77	83.33	100.00	88.89	90.91
0.55	93.75	76.92	97.44	96.55	90.91
0.65	96.97	66.67	97.44	96.55	83.33
0.75	96.97	50.00	97.44	96.77	80.00
0.85	91.43	40.00	97.44	90.91	66.67
0.95	91.43	40.00	97.44	88.24	54.55

Table 9 Accuracy

Threshold	Doc#1	Doc#2	Doc#3	Doc#4	Doc#5
0.05	35	80	20	30	80
0.15	70	90	40	50	85
0.25	85	95	70	65	90
0.30	95	95	75	70	90
0.40	95	90	90	80	95
0.50	95	90	100	85	95
0.55	90	85	95	95	95
0.65	95	75	95	95	90
0.75	95	50	95	95	85
0.85	85	25	95	85	70
0.95	85	25	95	80	50

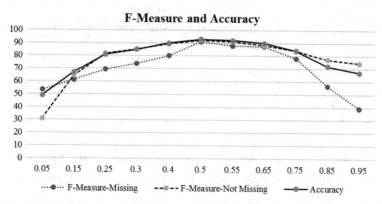

Fig. 3 Average F-measure and accuracy with regard to all SRB documents

and so the tool uses the threshold between 0.4 and 0.5. Note that, at the threshold of 0.55, the F-measure of the Missing class for Doc#3 is not applicable (N/A) because there is no SSRB with the difference score of at least 0.55 and predicted as Missing. So the precision, and hence the F-measure, of the Missing class cannot be calculated.

On taking a closer look at Doc#2 and Doc#5, we notice that the performance is particularly high even at the low threshold of 0.05–0.25, compared with the other three documents. As the SSRBs usually mention technical security terms as the examples of the techniques that should be implemented, Doc#2 and Doc#5 which are written in more technical terms, match better with the SSRBs than the other three documents which are written in more general terms and do not contain many technical terms. For example, an SSRB states that *"The system shall encrypt confidential information (e.g. user ID, password encryption key, database user id, database password) by strong encryption (e.g. AES 128 bits, AES 256 bits, RSA*

2048)." While Doc#1 is written as "*The application shall apply strong encryption to encrypt confidential information and store in a database or configuration file with appropriate access control.*", Doc#5 is written as "*The application shall use AES 256 bits encryption to encrypt system user id, password, database user id, and password before storing those hashes in a configuration file.*" The difference score of Doc#5 with more technical term is lower, and this goes along with the view of the auditors who are less likely to agree with a requirement as a general statement. The auditors usually expect the banks to be explicit about the techniques used whenever possible.

We further experiment by dividing the SRB documents into two groups, i.e. Doc#1, Doc#3, and Doc#4 that use less technical terms and Doc#2 and Doc#5 that use more technical terms. The performance with regard to the two groups is in Figs. 4 and 5. For the non-technical terms group, the best performance is still when the threshold is 0.5 as there are false negatives when the threshold is higher. For the

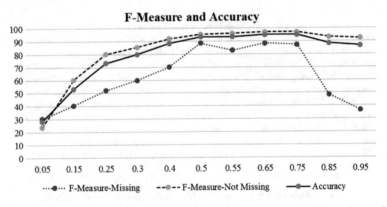

Fig. 4 Average F-measure and accuracy with regard to SRB documents that use less technical terms

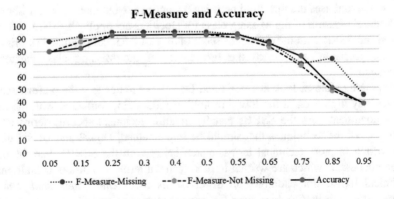

Fig. 5 Average F-measure and accuracy with regard to SRB documents that use more technical terms

technical terms group, the performance is best even at the low threshold of 0.25 and continues so until the threshold reaches 0.5 where the performance subsequently drops and false negatives appear. This experiment suggests the requirements engineers who use the tool to consider the writing style of the SRB document and may adjust the threshold accordingly.

6.2 Validity of Risk Index

To validate the risk index, we use Spearman's rank order correlation to determine the correlation between the risk index of each SRB document as calculated by the tool and the ordinal risk level determined by 12 security engineers and network engineers as in Table 10. We map the ordinal risk level of (Very Low, Low, Medium, High, Very High) given by the engineers to a numerical scale of 1–5 respectively. For the risk index by the tool whose value is in [0, 25], we map the range [0, 5] to 1, (5, 10] to 2, (10, 15] to 3, (15, 20] to 4, and (20, 25] to 5.

The hypotheses are

H_0:   There is no monotonic correlation between the risk level by the engineers and the risk index by the tool ($\rho_s = 0$).

H_1:   There is monotonic correlation between the risk level by the engineers and the risk index by the tool ($\rho_s \neq 0$).

The calculated correlation coefficient $r_s = 0.89 \approx 0.9$. Since r_s is not less than $r_{critical} = 0.9$ at the significance level $\alpha = 0.1$ and df = 5, we reject H_0 and accept H_1. There is monotonic correlation between the risk level by the engineers and the risk index by the tool at $\alpha = 0.1$. The correlation is strong and positive.

To experiment further, we consider the total of 42 missing SSRBs from all five SRB documents, together with their associated risk indices, to test monotic correlation with the risk levels given by the engineers. The distribution of the mapped risk levels of all 42 SSRBs is shown in Fig. 6. In this case, since $r_s = 0.74$ is not less than $r_{critical} = 0.305$ at the significance level $\alpha = 0.05$ and df = 42, we again reject H_0 and accept H_1. For the case of missing SSRBs, there is also monotonic correlation between the risk level by the engineers and the risk index by the tool at $\alpha = 0.05$. Again, the correlation is quite strong and positive.

Table 10 Risk level and risk index of each SRB document using threshold of 0.5

	Risk level by engineers	Mapped Risk Level by engineers	Risk Index by Tool	Mapped risk level by tool
Doc#1	Medium	3	8.96	2
Doc#2	Very high	5	14.62	3
Doc#3	Low	2	4.39	1
Doc#4	Very high	5	20.34	5
Doc#5	Very high	5	19.8	4

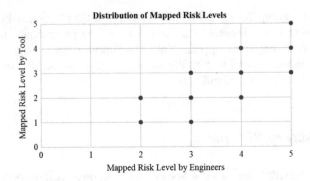

Fig. 6 Distribution of mapped risk levels

7 Conclusion

Given its satisfactory performance, the proposed automated risk assessment of security requirements of banking information systems is a useful approach to supporting requirements engineers or business analysts to check for compliance with security regulations. The approach can help point out where the weaknesses are in the security requirements document and prioritize the improvement. Even though the context of this paper is the banking sector, we believe the approach can be applied to other domains where checking requirements for compliance with regulations and attack-based security measurement are desirable.

To further improve the performance, we plan to experiment on using domain-specific stop words, semantics-related words, and spelling correction prior to the analysis. Human correction can also be allowed to adjust compliance checking and risk index results, e.g. based on specific environmental setting.

References

1. Security Scorecard HQ: 2016 Financial Industry Cybersecurity Report: SecurityScore card. https://cdn2.hubspot.net/hubfs/533449/SecurityScorecard_2016_Financial_Report.pdf (2016). Accessed 20 May 2017
2. Bank of Thailand: IT Best Practices Phase I, Thailand (2013)
3. Bank of Thailand: IT Best Practices Phase II, Thailand (2014)
4. The MITRE Corporation: CAPEC-Common Attack Pattern Enumeration and Classification. http://capec.mitre.org. Accessed 15 April 2017
5. Li, T., Paja, E., Mylopoulos, J., Horkoff, J., Beckers, K.: Security attack analysis using attack patterns. In: Proceeding of 2016 IEEE 10th International Conference on Research Challenges in Information Science (RCIS), pp. 1–13 (2016). doi:10.1109/RCIS.2016.7549303
6. Steirna, E.J., Rowe, N.C.: Applying information-retrieval methods to software reuse: a case study. J. Inf. Process. Manage. **39**(1), 67–74 (2003). doi:10.1016/S0306-4573(02)00025-0
7. Baeza-Yates, R., Ribeiro-Neto, B.L: Modern Information Retrieval, 2nd ed. ACM Press, New York (2011)

8. Ilyas, M., Kung, J.: A similarity measurement framework for requirements engineering. In: Proceeding of 2009 Fourth International Multi-Conference on Computing in the Global Information Technology, Cannes, La Bocca, pp. 31–34 (2009). doi:10.1109/ICCGI.2009.12

9. Dag, J.N.O., Regnell, B., Carlshamre, P., Andersson, M., Karlsson, J.: A feasibility study of automated natural language requirements analysis in market-driven development. J. Requirements Eng. **7**, 20 (2002). doi:10.1007/s007660200002

10. Yu, Y., Franqueira, V.N.L., Tun, T.T., Wieringa, R.J., Nuseibeh, B.: Automated analysis of security requirements through risk-based argumentation. J. Syst. Softw. **106**, 102–116 (2015). doi:10.1016/j.jss.2015.04.065

11. Piromsopa, K., Rojkangsadan, T., Prompoon, N.: A Risk assessment of web server Impact classification by loss type. In: Proceeding of Networks and Communication Systems (NCS), pp. 173–178 (2005)

12. Banklongsi, T., Senivongse, T.: A security measurement model for web services based on provision of attack countermeasure. In: Proceeding of 15th International Annual Symposium Computational Science and Engineering (ANSCSE15), pp. 593–598 (2011)

13. Firesmith, Donald G.: Engineering security requirements. J. Object Technol. **2754**, 53–68 (2003)

mCITYPASS: Privacy-Preserving Secure Access to Federated Touristic Services with Mobile Devices

Macià Mut-Puigserver, M. Magdalena Payeras-Capellà, Jordi Castellà-Roca and Llorenç Huguet-Rotger

Abstract Destination cards are offered in many cities to provide tourists a simple and integrated way to access interest points and public transport. In the one hand, these systems, called citypasses, are usually inefficiently implemented on smartcards. In the other hand, existing e-ticketing proposals are general purpose systems or created for its application to transport services and they are not adapted to the requirements of a citypass system. In this paper we present an electronic ticketing system intended to be used for touristic services. Our proposal is the first one that can be implemented on portable devices, such as smartphones, and is flexible enough to include reusable and non-reusable services in the same citypass. The system has been designed taking into account all the security and privacy requirements described for electronic tickets, including the challenging ones (exculpability, reusability and unsplittability) obtaining a very secure and powerful system. The dispute resolution protocol assures that all the parts are protected against other part's attacks. Finally, the system allows the user to use the system anonymously, so the privacy of the system is assured.

Keywords Secure access · Tourism · e-ticketing · e-commerce · Security · Privacy · Exculpability · Smart cities · Mobile applications

M. Mut-Puigserver · M. Magdalena Payeras-Capellà (✉) · L. Huguet-Rotger
Department de C. Matemàtiques i Informàtica, Universitat de les Illes Balears,
Ctra. de Valldemossa, km 7,5, 07122 Palma, Spain
e-mail: mpayeras@uib.cat

M. Mut-Puigserver
e-mail: macia.mut@uib.cat

L. Huguet-Rotger
e-mail: l.huguet@uib.cat

J. Castellà-Roca
Department d'Enginyeria Informàtica i Matemàtiques, UNESCO Chair in Data Privacy,
Universitat Rovira i Virgili. Av. Països Catalans 26, 43007 Tarragona, Spain
e-mail: jordi.castella@urv.cat

© Springer International Publishing AG 2018
R. Lee (ed.), *Applied Computing & Information Technology*,
Studies in Computational Intelligence 727, DOI 10.1007/978-3-319-64051-8_9

1 Introduction

Tourism is subject to a transformation due to the use of communication technology. Information and communication systems are enabling an integrative technology in nearly all areas of daily life and, as might be expected, they also affect urban transport and tourism. Nowadays, almost all big cities and touristic areas have created new digital touristic tickets to replace the old paper-based booklets containing tickets for touristic attractions. Thus, information technologies offer integrated services to promote a digital access the touristic sites (museums, monuments, attractions, exhibitions,…) and public transport. That is the motivation of this paper.

The integrated e-ticketing schemes are often called *destination cards* or *city passes* that are valid for a specific period (lifetime). Moreover, as additional advantages, the majority of these e-ticketing schemes habitually offer discounts in complementary services as well as direct access to the sites, so users can skip queues [1]. Usually, destination cards are implemented over smartcards (activated before its first use), or paper booklets including bar codes. Examples of these cards can be found in *"We love city cards"* [2], a portal to destination card systems of 36 European cities or CityPASS [3] a portal to 12 American cities. The most relevant advantages are: free entry to top visitor attractions, offers and discounts, priority access, choice of cards for different periods, free guidebook, city maps and limited or unlimited access to public transport.

We propose a new model for destination cards based on the use of portable devices, such as smartphones. The idea is to store the booklet of e-tickets on mobile devices. The advantages of this approach in front of smartcard-based schemes is that the user can electronically buy online his electronic booklet in advance and store the e-tickets on it (avoiding queues at the sale counter) and no additional devices are needed (users are expected to have smartphones[1]). At the provider side, the smartphone approach avoids the reader-recorder that is needed for smartcards and reduces expenditure management through online counters.

In this paper we introduce a new protocol for electronic touristic tickets named *mCITYPASS* that allows the purchase and use of a booklet of multiple access tickets to touristic attractions and public transport. Depending on the services included in the booklet, these e-tickets can be non-reusable, *m*-times reusable or infinitely reusable. In real scenarios, non-reusable tickets are usually related to attractions while reusable tickets are usually related to transport, but the specific use depends on the city. As an example, *Granada CityPASS* (lovegranada.com) offers a 3-day pass that includes *The Alhambra* and *Generalife* entrances and 5 urban bus journeys while the pass for five days includes *The Alhambra* and *Generalife* entrances, 9 urban bus journeys and 1 tourist train trip.

---

[1]The market penetration of smartphones in 2015 was in South Korea at 88%, Australia (77%), Israel (74%), USA (72%) and Spain (71%) according to surveys conducted by Pew Research Center (http://www.pewresearch.org).

1.1 Contribution

The use of e-ticketing systems for transport services has been largely studied. We presented a survey on E-ticketing applied to transport in [4]. These survey detailed all the desirable properties and requirements for an e-ticketing system. Later, we proposed an e-ticketing scheme fulfilling the requirements of exculpability and reusability in [5]. However, in order to address the challenge of a mobile application for a citypass system, we have to create a specific e-ticketing system that fulfils the particularities of this touristic application.

The contribution of this paper is a new secure system that provides reusability of some services and allows not only the secure access to transportation systems but also to other touristic services like museums, attractions, etc. Some of the services included in the citypass can only be accessed once while others can be accessed multiple times or even infinitely. The system also protects the privacy of users and avoids the generation of identified user profiles. In addition, it must be secure enough to avoid forgery. Finally, we have designed the system as a fair scenario for both users and service providers[2] so all the parts are protected against other part's attacks or denial of service.

Current solutions based on smartcards rely on the physical security of the token but require a framework for the generation and the management of the cards. An implementation on mobile devices would make the system more convenient and easier to access for tourists.

The paper proposes a system for a secure and fair citypass system implemented on mobile devices. The main contribution of the proposal is a system that is completely adapted to the particularities of touristic services and fulfils the required properties, not only the more usual ones but the more challenging, like exculpability, unsplittability and selective reusability. These properties are achieved without the need to use any tamper proof device, like smartcards.

1.2 Organization

This paper is organized as follows. Section 2 reviews the state of the art, including security requirements and related proposals. Section 3 describes in detail all the phases of the protocol while Sect. 4 includes the claims that guarantee the fairness and the exculpability of the scheme. Section 5 includes some facts about the future deployment and analysis of the system and finally Sect. 6 lists the conclusions of the paper. From now on, we will use the term citypass to refer to a system that manages destination cards or citypasses, mcitypass to refer to our proposal and pass to refer to a single citypass issued by the system.

---

[2]A fair exchange protocol occurs when, at the end of any protocol run, any honest participant has received all expected items or none of them has obtained any valid item.

2   State of the Art

In this section we will describe the state of the art in destination cards. First, the security requirements for this kind of applications are listed and then the most significant proposals are described and evaluated.

2.1   Security Requirements

The destination card systems have to consider and guarantee the following security requirements, described in [4] and adapted to the particularities of the destination cards:

- Authenticity: A user has to be able to verify if a PASS has been issued by an authorized issuer.
- Integrity: All the parts have to be able to verify if the PASS has been altered as regards to the one issued by the correspondent authorized issuer.
- Non-repudiation: Once a valid PASS has been issued, the issuer has not to be able to deny that he has issued that PASS.
- Unforgeability: Only authorized issuers can issue a valid PASS.
- Non-Overspending and reusability: The PASS will include several tickets, both non-reusable and reusable. In both cases, ticket overspending has to be prevented. Tickets included in the PASS can only be used as agreed between the issuer and the user. Non-reusable tickets can not be reused after they have been spent. Reusable tickets can be used exactly the number of times agreed in the moment of issue.
- Revocable anonymity: The scheme allows the revocation of the anonymity of the user if he misbehaves using the service, otherwise the user remains anonymous.
- Expiration date and lifetime: A PASS will be only valid during a time interval.
- Fairness: During the execution of the protocol the parties execute several exchanges of elements, this exchanges could not lead to unfair situations.
- Exculpability: The provider can not falsely accuse the user of ticket overspending, and the user is able to demonstrate that he has already validated the ticket before using the service.
- Unsplittability: Users should not be able to share a PASS in such a way that several users validate tickets from the same PASS.

2.2   Related Proposals

There are some papers that explore the idea of the development of integrated e-ticketing systems for touristic sites in cities. The idea behind these tickets is to combine several touristic services (e.g. transport passes, leisure activities tickets, tourist attractions vouchers) on a single ticket. The integrated ticket has the advantage of having a lower fare than buying the passes of the different services separately. For example, these systems became a goal of the European Union applied to the transports systems in cities [6]. Though, e-ticketing systems must be secure and have to introduce measures to avoid counterfeit copies and preserve user's privacy.

Many large cities have introduced multimodal e-ticket systems in public trans-portation and tourist attractions [2, 3]. The technology used in them may vary from one city to another (e.g. contactless or contact-based smartcards). Nevertheless, the use of smartphones remains residual. In Helsinki, Finland, public transport users have been able to pay their fares with their mobile phones since 2001 [7]. However, in some cases (e.g. New York CityPASS), vouchers must be printed and exchanged for the user's CityPASS booklet at the visited attraction. In [8], the authors explain the practical advantages of operating with mobile devices, which reduces the cost of system infrastructure.

Also, there are some papers [5, 9–11] that explore the advantages and the development of e-ticketing schemes in touristic environments and transport systems. Nevertheless, just a few of them have some specification on how to implement such systems. Alessadra et al. in [12] describes an application for Android smartphones offering, among others, mobile ticketing services. However, the paper only presents a generic framework and does not describe any e-ticketing implementation.

In some papers mobile phones are used as virtual vouchers, e-tickets applied to the transport system, or even supermarket loyalty cards. Sang-Won et al. in [13] suggests a NFC-based mobile ticket for small traders and enterprisers that can be used in touristic applications. However, this proposal does not store the ticket on the smartphone, they just use the mobile device to identify the user, thus anonymity is not possible in such schemes. In [14] the author proposes the so called *mCoupons* stored in the mobile device. The scheme proposed in this paper is only applied to single tickets and the security of the systems only deals with against multiple spending, unauthorized generation and manipulation, and copying. Han-Cheng in [15] proposes a similar scheme (same features than [14] but a more efficient implementation).

Multi-ticketing systems were introduced to increase the efficiency of having tickets separately [16]. Chen et al. [17] introduced a multi-ticketing system, which provides privacy-protection and also protection against splitting. However, the tickets contained in the same multi-ticket can only be redeemed in the order that was fixed during the issue protocol. Some inefficient issues of the Chen's protocol were improved in [16]. In any case, only a single vendor is considered.

In order to make more appealing to users, multi-ticketing systems with a fed-eration of vendors has been designed in [18]. In this case, the mobile device can store a multi-ticket token that contains single-tickets from a federation of vendors (e.g. a corporation of touristic attractions managed by different vendors from the same city). This proposal presents unsplittability but reusability, exculpability, activation and fairness are not contemplated in the system.

3 mCITYPASS Scheme

The scheme has the following actors: the user $\mathcal{U}$; the ticket issuer $\mathcal{I}$, who sends a valid ticket to $\mathcal{U}$; a set of service providers P_i, who verify the tickets inside the PASS and give the corresponding service; and finally a trusted third party (TTP) $\mathcal{T}$, who preserves $\mathcal{U}$'s anonymity, and gives a valid non-identity pseudonym to $\mathcal{U}$. The mCITYPASS (mobile CITYPASS) scheme has been designed for mobile devices (smartphones, tablets, smartwatches,…), reducing the computation requirements in the user side, and providing the basic security requirements (authenticity, non-repudiation and integrity) together with expiry and activation date, revocable anonymity, exculpability, reusability and unsplittability.

In Tables 1 and 2 we define the notation used in the description of our scheme.

3.1 Phases

The phases of our system are: *Providers' Affiliation*, that includes the procedure to affiliate each Service Provider and the procedure to generate the parameters of the mCITYPASS scheme; *Pseudonym*, where the user obtains a new temporary pseudonym to be used in the system without linkage to user's identity (if user behaves correctly); *PASS Purchase*, that consists on the payment and reception of the PASS; the *PASS Activation* that can be executed together with the *PASS Purchase* o later, and *PASS Verification*, where the user shows a ticket included in the PASS to a service provider in order to be checked and validated. Other phases considered in the system are claims, that should only be executed in case of controversial situations during the *PASS Verification* phase.

Table 1 Cryptographic primitives used in the protocol description

Cryptography: Notation and Description	
$sk_{\mathcal{E}}(content)$	Decryption of *content* or the generation of a signature with its *content* by using the private key of the entity $\mathcal{E}$
$pk_{\mathcal{E}}(content)$	Encryption of *content* or the verification of a signature *content* by using the public key of the entity $\mathcal{E}$
hash()	Public cryptographic one-way summarizing function that achieves collision-resistance
$hash^m()$	Represents the hash function applied m times

Table 2 Data structures and items used in the protocol description

INFORMATION ITEMS: Notation and Description			
$Cert_{\mathscr{U}}$	User's Digital Credential	$Pseu_{\mathscr{U}}$	User Temporary pseudonym
$Cert_P[1 \ldots J]$	Array of all the providers' certifications	Sn	Serial number of the PASS
γ_i	Index for a non-reusable service	λ_i	Index for a reusable service
ξ_i	Index for an infinitely reusable service	K	Shared session key
RI_i	Secret random value sent by P_i to $\mathscr{U}$ in order to give her the right to use a service	RU	Secret random number to demonstrate the ownership of a PASS
$\delta_{\mathscr{T},P_i}$	Digital envelope of κ_i	$\delta_{\mathscr{T},\mathscr{P}}[\ ]$	Array of all digital envelopes for all services in a PASS
$h_{RI}[\ ]$	Array with all hashes on all secret random numbers RI_i	$H_{\mathscr{U}}$	Hidden RU
τ_1	First verification timestamp. P_i generates it during the verification process of a PASS at *verifyPASS* stage	τ_2	Second verfication timestamp. P_i creates it during the verification process of a PASS at *verifyProof* stage
$A_{\mathscr{P}}$	Vernam cipher of RI_i	$A_{\mathscr{U}}$	Vernam cipher of $\psi_{i,j}$
k_{λ_i}	Counter to keep track of the number of times that the ticket can be used	ACT	Activation proof of a PASS
$\psi_{i,0}$	Secret value, the knowlege of this value by $\mathscr{U}$ proves the right to use a non-reusable service	$\psi_{i,j}$	Value derived from $\psi_{i,0}$ using j-times a hash function. The knowlege of this value by $\mathscr{U}$ proves the right to use a reusable service
V_{succ}	Message generated by P_i, the meaning of this message is denoted by flag$_1$ or flag$_2$	V_{fail}	Message generated by P_i, the meaning of this message is denoted by flag$_0$
κ_i	Information generated by $\mathscr{I}$ to P_i and $\mathscr{T}$ to give $\mathscr{U}$ the right to use a service	flag$_{10}$	Indicates that the ZKP has finished successfully. So, the service can still be used by $\mathscr{U}$ using the PASS. In case of a reusable service, it also specify that P_i agrees with counter k_{λ_i}
flag$_{00}$	It is used by P_i to indicates whether any parameter of the PASS sent by $\mathscr{U}$ has not the proper form (e.g. the current date is past the final expiry date) or its signature are not correct	flag$_{01}$	Indicates the problem: ZKP ends unsuccessfully ($\mathscr{U}$ has not demonstrate the ownership of the PASS)

(continued)

Table 2 (continued)

INFORMATION ITEMS: Notation and Description			
flag_{02}	Indicates that the ticket has been spent, i.e. is not valid anymore	flag_{03}	Specifies two possible errors: $\psi_{\gamma_i,0}$ does not match with $\psi_{\gamma_i,1}$, or τ_2 does not match the expiry date of the PASS
flag_{04}	Indicates that the ticket is not valid or some of the parameters sent by $\mathcal{U}$ are not correct	flag_{05}	It specifies that $\psi_{\lambda,(k_{\lambda_i}-1)}$ received in the previous step is not correct or τ_2 does not match the expiry date of the PASS

Users have a digital credential ($\text{Cert}_{\mathcal{U}}$) only for authentication to the TTP, since the system is anonymous, and all further movements in the system are tracked only with the assigned temporary pseudonym ($\text{Pseu}_{\mathcal{U}}$).

3.1.1 PHASE 1: Providers' Affiliation

The first stage of the scheme is the affiliation of the service providers. They have to contact with the issuer $\mathcal{I}$ and enroll as a provider of a certain service in the framework of the mCITYPASS. The number of services included in the PASS are denoted by J, so J providers are going to be joined to the scheme. As a result, $\mathcal{I}$ will have an array where all public key certificates of the providers are arranged. A provider P_i can supply either a non-reusable service (γ_i), a m-times reusable service (λ_i) (this means that each reusable service λ_i has a maximum number of uses identified by m) or a infinitely reusable service (ξ_i). The protocol is as follows:

$\mathcal{I}$ chooses a certain public key algorithm (e.g., RSA, DSA) as a reference for the creation of P_i's public-key pairs.

authenticateProvider Provider P_i follows the next steps:

1. generates a public-key pair for the chosen public key algorithm and the appropriate parameters;
2. sends this cryptographic information and the information of its service (either γ_i, λ_i or ξ_i depending on the reusability of the service) to $\mathcal{I}$ via an authenticated channel;

generateCertificate Issuer $\mathcal{I}$ executes:

1. verify that the information is correct;
2. generate a public key certificate for P_i with the key usage (i.e., linked to the service γ_i or λ_i);

Fig. 1 Example of the seeds included in a PASS with 7 services: one infinitely reusable service, three non-reusable services, two four-time reusable services and one twice-reusable service

When all J providers have executed the previous protocol the issuer can generate the array of certificates:

arrayGeneration Issuer $\mathscr{I}$ executes:

1. order and arrange the series of P_i's certificates in an array: $Cert_P[1...J]$;

   After the affiliation of P_i, the structure of the **PASS** can be created (see Fig. 1).

3.1.2 PHASE 2: Pseudonym

User $\mathscr{U}$ contacts the pseudonym manager $\mathscr{T}$ in order to obtain the assigned pseudonym. The certificate $\mathsf{Cert}_{\mathscr{U}}$ identifies $\mathscr{U}$ through a secure connection established between the two parties. The system has cryptographic public parameters [19], which have been published: (α, p, q), where α is a generator of the group G with order p, being p and q large primes achieving $p = 2q + 1$.

$\mathscr{U}$ generates a random value $\mathsf{x}_{\mathscr{U}} \xleftarrow{R} \mathbb{Z}_q$ and computes $\mathsf{y}_{\mathscr{U}} = \alpha^{\mathsf{x}_{\mathscr{U}}} \pmod{p}$ in order to receive a valid signed pseudonym $\mathsf{Pseu}_{\mathscr{U}}$ from $\mathscr{T}$. $\mathscr{U}$ and $\mathscr{T}$ have their own pair of keys used for signature and encryption of the transmitted data between them. The protocol is as follows:

authenticateUser User $\mathscr{U}$ follows the next steps:

1. generates $\mathsf{x}_{\mathscr{U}} \xleftarrow{R} \mathbb{Z}_q$, and computes $\mathsf{y}_{\mathscr{U}} = \alpha^{\mathsf{x}_{\mathscr{U}}} \pmod{p}$;
2. computes the signature $\mathsf{sk}_{\mathscr{U}}(\mathsf{h}_{\mathsf{y}_{\mathscr{U}}})$ where $\mathsf{h}_{\mathsf{y}_{\mathscr{U}}} = hash(\mathsf{y}_{\mathscr{U}})$;
3. encrypts $\mathsf{y}_{\mathscr{U}}$ with $\mathscr{T}$'s public key: $\mathsf{pk}_{\mathscr{T}}(\mathsf{y}_{\mathscr{U}})$;
4. sends $(\mathsf{sk}_{\mathscr{U}}(\mathsf{h}_{\mathsf{y}_{\mathscr{U}}}), \mathsf{Cert}_{\mathscr{U}}, \mathsf{pk}_{\mathscr{T}}(\mathsf{y}_{\mathscr{U}}))$ to $\mathscr{T}$;

generatePseudonym Pseudonym Manager $\mathscr{T}$ executes:

1. decrypts $\mathsf{sk}_{\mathscr{T}}(\mathsf{pk}_{\mathscr{T}}(\mathsf{y}_{\mathscr{U}})) \rightarrow (\mathsf{y}_{\mathscr{U}})$;
2. verifies $\mathsf{y}_{\mathscr{U}}$: $\mathsf{pk}_{\mathscr{U}}(\mathsf{sk}_{\mathscr{U}}(\mathsf{h}_{\mathsf{y}_{\mathscr{U}}})) \rightarrow (\mathsf{h}_{\mathsf{y}_{\mathscr{U}}}) \overset{?}{=} hash(\mathsf{y}_{\mathscr{U}})$;
3. if correct, then computes the signature of $\mathsf{sk}_{\mathscr{T}}(\mathsf{h}_{\mathsf{y}_{\mathscr{U}}})$; and
4. sends $\mathsf{Pseu}_{\mathscr{U}} = (\mathsf{y}_{\mathscr{U}}, \mathsf{sk}_{\mathscr{T}}(\mathsf{h}_{\mathsf{y}_{\mathscr{U}}}))$ to $\mathscr{U}$.

verifyPseudonym $\mathscr{U}$ computes:

1. verifies $y_{\mathscr{U}}$: $\mathsf{pk}_{\mathscr{T}}(\mathsf{sk}_{\mathscr{T}}(h_{y_{\mathscr{U}}})) \to (h_{y_{\mathscr{U}}}) \stackrel{?}{=} hash(y_{\mathscr{U}})$;

If in the future the user visits the same city and wants to buy another PASS, he can execute this phase again in order to avoid *linkability* in the usage of the anonymous tickets.

3.1.3 PHASE 3: PASS Purchase

The user establishes a connection with the PASS issuer $\mathscr{I}$ in order to receive the pass. This connection could be established through an anonymous channel like TOR [20], guaranteeing user's privacy. There are current contributions[3] that have implemented TOR for Android devices. $\mathscr{I}$ has a key pair and a public key certificate ($\mathsf{Cert}_{\mathscr{I}}$). Users do not use their personal keys (it would cause loss of anonymity); instead they use the temporal pseudonym and authenticate through the Schnorr's Zero-Knowledge Proof (ZKP) [21]. The payment method is considered as out of scope in this proposal as we focus on the privacy given to the user when joining the system, and using the services.

In this phase, the user selects a PASS Type, that defines the duration of the PASS, Lifetime, and the Category of the user (adult, child, young,...): Type(Lifetime, Category). $\mathscr{I}$ generates the PASS with all the required information and its digital signature, together with an array of J secret values RI_i (where $i = 1...J$, let J be the number of providers affiliated to the PASS service) and the set of secret shared keys (they are decryptable only by the corresponding provider P_i and $\mathscr{T}$) in order to let each provider show the secret value RI_i related to each provider later, in the verification phase. The ticket issuer $\mathscr{I}$ and the user $\mathscr{U}$ follow this protocol:

getService $\mathscr{U}$ executes:

1. selects the desired Type of PASS Type(Lifetime, Category);
2. generates a of random value $\mathsf{RU} \stackrel{R}{\leftarrow} \mathbb{Z}_q$;
3. computes $\mathsf{H}_{\mathscr{U}} = \alpha^{\mathsf{RU}} \pmod{p}$;
4. generates two more random values $a_1, a_2 \stackrel{R}{\leftarrow} \mathbb{Z}_q$ to be used in the Schnorr proof;
5. computes $A_1 = \alpha^{a_1} \pmod{p}$;
6. computes $A_2 = \alpha^{a_2} \pmod{p}$;
7. sends $(\mathsf{Pseu}_{\mathscr{U}}, \mathsf{H}_{\mathscr{U}}, A_1, A_2, \mathsf{Type})$ to the ticket issuer $\mathscr{I}$.

---

[3]http://sourceforge.net/apps/trac/silvertunnel/wiki/TorJavaOverview.

getChallenge $\mathcal{I}$ follows the next steps:

1. generates and sends a challenge $\mathsf{c} \xleftarrow{R} \mathbb{Z}_q$ for $\mathcal{U}$;
2. asynchronously, for optimization, pre-computes $\mathsf{y}_{\mathcal{U}}^{\mathsf{c}}$ $(\mathrm{mod}\, p)$ and $\mathsf{H}_{\mathcal{U}}^{\mathsf{c}}$ $(\mathrm{mod}\, p)$;

solveChallenge $\mathcal{U}$ computes:

1. computes $\mathsf{w}_1 = a_1 + \mathsf{c} \cdot \mathsf{x}_{\mathcal{U}}$ $(\mathrm{mod}\, q)$;
2. computes $\mathsf{w}_2 = a_2 + \mathsf{c} \cdot \mathsf{RU}$ $(\mathrm{mod}\, q)$;
3. pre-computes the shared session key used in the ticket verification: $\mathsf{K} = hash(\mathsf{w}_2)$;
4. generates a set of J randoms, let J be the number of providers offering a service in the **PASS**, $\psi_{1,0}, \ldots, \psi_{J,0} \xleftarrow{R} \mathbb{Z}_q$
5. for each non-reusable service γ_i, calculates $\psi_{\gamma_i,1} = hash(\psi_{\gamma_i,0})$
6. for each reusable service λ_i (suppose that m is the number of uses of service λ_i), computes $\psi_{\lambda_i,m} = hash^m(\psi_{\lambda_i,0})$ where $hash^m()$ represents the hash function applied m times; see Fig. 2
7. stores $\psi_{1,0}, \ldots, \psi_{J,0}$ values in his *TicketsPASS* database;
8. encrypts and sends the generated information to $\mathcal{I}$: $\mathsf{pk}_{\mathcal{I}}((\mathsf{w}_1, \mathsf{w}_2), (\{\psi_{\gamma_i,1}, \psi_{\lambda_i,m}\} \forall \gamma_i, \lambda_i))$ and pays for the **PASS**;

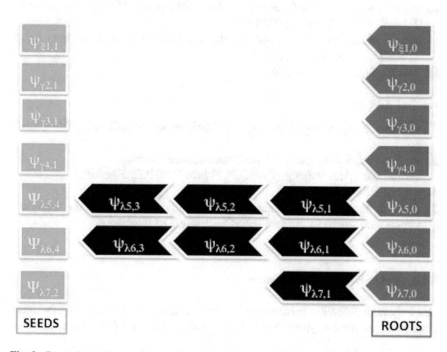

Fig. 2 Generation of the seed for an Example of a PASS with 7 services: one infinitely reusable service, three non-reusable services, two four-time reusable services and one twice-reusable service

getPASS $\mathscr{I}$ follows the next steps:

1. decrypts $\mathsf{sk}_{\mathscr{I}}(\mathsf{pk}_{\mathscr{I}}(\mathsf{w}_1, \mathsf{w}_2)) \rightarrow (\mathsf{w}_1, \mathsf{w}_2)$;
2. computes $\alpha^{\mathsf{w}_1} \ (\mathrm{mod}\, p)$;
3. computes $\alpha^{\mathsf{w}_2} \ (\mathrm{mod}\, p)$;
4. verifies $\alpha^{\mathsf{w}_1} \overset{?}{=} A_1 \cdot \mathsf{y}_{\mathscr{U}}^{\mathsf{c}} \ (\mathrm{mod}\, p)$;
5. verifies $\alpha^{\mathsf{w}_2} \overset{?}{=} A_1 \cdot \mathsf{H}_{\mathscr{U}}^{\mathsf{c}} \ (\mathrm{mod}\, p)$;
6. if both verifications hold, $\mathscr{I}$ has checked that $\mathscr{U}$ is an authorized user. Otherwise, $\mathscr{I}$ stops the protocol;
7. computes the shared session key: $\mathsf{K} = hash(\mathsf{w}_2)$;
8. obtains a unique serial number Sn, and a set of random values $\mathsf{RI}_i \overset{R}{\leftarrow} \mathbb{Z}_p$, for $i = 0, \ldots, J$;
9. computes the set $\mathsf{h}_{\mathsf{RI}_i} = hash(\mathsf{RI}_i)$ for $i = 1, \ldots, J$;
10. composes the set $\kappa_i = (\mathsf{K}, \mathsf{RI}_i)$ and signs it $\kappa_i^* = (\kappa_i, \mathsf{Sign}_{\mathscr{I}}(\kappa_i)) \forall i = 1, \ldots, J$;
11. encrypts each κ_i^* with a digital envelope which is decryptable by the TTP $\mathscr{T}$ and the provider P_i for possible future controversial situations during the ticket verification: $\delta_{\mathscr{T}, P_i} = \mathsf{pk}_{\mathscr{T}, \mathscr{P}_i}(\kappa_i^*)$.
12. fills out the PASS information $\mathsf{PASS} = (\mathsf{Sn}, \mathsf{Type}, \mathsf{Pseu}_{\mathscr{U}}, \mathsf{Lifetime}, \mathsf{PURdate}, \mathsf{EXPdate}, \mathsf{h}_{\mathsf{RI}}[\ ], \mathsf{H}_{\mathscr{U}}, \delta_{\mathscr{T}, \mathscr{P}}[\ ], \psi_{\lambda, m}[\ ], \psi_{\gamma, 1}[\ ], \mathsf{Terms\ and\ Conditions})$;
13. digitally signs the PASS, $\mathsf{Sign}_{\mathscr{I}}(\mathsf{PASS}) = \mathsf{sk}_{\mathscr{I}}(hash(\mathsf{PASS}))$, and generates $\mathsf{PASS}^* = (\mathsf{PASS}, \mathsf{Sign}_{\mathscr{I}}(\mathsf{PASS}))$;
14. stores in its *CityPASS* database the information related to this ticket: $(\mathsf{PASS}^*, \kappa_i \ \forall i = 1, \ldots, J)$
15. sends PASS^* to the user $\mathscr{U}$.

receivePASS $\mathscr{U}$ executes:

1. verifies the digital signature $\mathsf{Sign}_{\mathscr{I}}(\mathsf{PASS})$ of the PASS using the issuer's certificate;
2. verifies PASS data and if the performed request match;
3. verifies the PASS validity ($\mathsf{PASS.PURdate}, \mathsf{PASS.EXPdate}$);
4. verifies $\mathsf{PASS.Pseu}_{\mathscr{U}}$;
5. stores in his *TicketsPASS* database $(\mathsf{PASS}^*, \mathsf{RU})$ together with the associated information stored in the step **solveChallenge**.7 of this phase.

3.1.4   PHASE 4: PASS Activation

One important difference between standard tickets and a CITYPASS is that whereas tickets can be used anytime before its expiration date, a PASS has a Lifetime

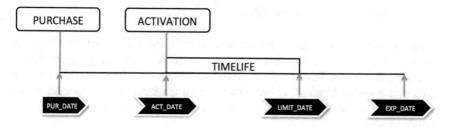

Fig. 3 PASS' Life Cycle

together with its expiration date, EXPdate. That means that the PASS must be used before its expiration date and will be valid during a Lifetime period [defined during the Pass purchase, Type(Lifetime, Category)]. This Lifetime period begins when the user activates the PASS. The activation can be performed immediately after the purchase or it can be performed just before the first use of the PASS.

We have described an activation between $\mathcal{U}$ and $\mathcal{I}$ but it can be performed between $\mathcal{U}$ and P_i.

Each PASS has a Purchase date, PURdate and an expiration date EXPdate fixed during the Pass Purchase phase. Moreover, each user selects the Type of card he wants to buy, including the validity period of the PASS, Lifetime. The activation phase sets the value of the activation date ACTdate. The limit date LIMdate is calculated adding the Lifetime to ACTdate. After LIMdate the services of the card cannot be used anymore. If the card is not activated before EXPdate or if LIMdate is greater than EXPdate, then the services cannot be used after EXPdate (see Fig. 3). The protocol is as follows:

showPass $\mathcal{U}$ computes:

1. sends PASS* to $\mathcal{I}$;

verifyTicket $\mathcal{I}$ executes:

1. verifies the PASS signature, PASS.PURdate, and PASS.EXPdate;
2. if the verifications fail, $\mathcal{I}$ aborts the ticket activation;
3. else $\mathcal{I}$ looks for the PASS PASS* in the database using PASS.Sn; and verifies that the PASS has not been activated: $\overset{?}{\exists}$ ACT linked to PASS* in the DB;

   a. if $\nexists$ ACT linked to PASS*:

      i. generates ACT = (PASS.Sn, ACTdate, *"Activated"*) and digitally signs ACT, and obtains the signed activation proof, $\text{Sign}_{\mathcal{I}}(\text{ACT}) = \text{sk}_{\mathcal{I}}(hash(\text{ACT}))$, and ACT$^* = (\text{ACT}, \text{Sign}_{\mathcal{I}}(\text{ACT}))$;
      ii. sends ACT* to the user $\mathcal{U}$.

3.1.5   PHASE 5: PASS Verification

When the user wants to access the service, he must show and verify the PASS in advance. The user only interacts with the service provider, but in controversial situations, he and/or the service provider could interact directly with the TTP through a *resilient channel*[4] in order to preserve the security requirements of the protocol. If the user misbehaved, his identity could be revoked, enabling to take further actions.

$\mathscr{U}$ sends the PASS*, and $\mathscr{P}$ checks it. If passed, $\mathscr{P}$ sends the commitment so that the corresponding RI will be disclosed if $\mathscr{U}$ behaves correctly. Once the user proves the knowledge of the value RU (related to the PASS) using a ZKP and reveals the secret value corresponding to the service, then he receives the secret RI together with the receipt R* from $\mathscr{P}$. We include the description of the validation of a non-reusable service and for a reusable service (prefixed with a number of uses m):

Verification of a non-reusable service γ_i, service provider P_i and user $\mathscr{U}$ follow these steps:

showPASS $\mathscr{U}$ computes:

1. generates a random value $a_3, \xleftarrow{R} \mathbb{Z}_q$ to be used in a Schnorr proof;
2. computes $A_3 = \alpha^{a_3} \pmod p$;
3. composes the information ticket message $m_1 = (\text{PASS}^*, \text{ACT}^*, A_3)$;
4. signs and sends it to P_i: $m_1^* = (m_1, \text{sk}_{\mathscr{P}i}(hash(m_1)))$;

verifyPASS P_i executes:

1. verifies the PASS signature, PASS.Sv, PASS.PURdate, and PASS.EXP date;
2. verifies ACT*, and PASS.ACTdate
3. calculates LIMdate = ACTdate + Lifetime
4. verifies that the present date is not greater that LIMdate
→ **if any verification fails**:
5. assigns $V_{\text{fail}} = (\text{PASS.Sn}, \text{flag}_{00}, \tau_1)$.
6. signs $V_{\text{fail}}^* = (V_{\text{fail}}, \text{sk}_{\mathscr{P}i}(hash(V_{\text{fail}})))$ and sends $m_2 = V_{\text{fail}}^*$ to $\mathscr{U}$
→ **else**:
5. looks for PASS* in its *SpentCityPASSES* database using PASS.Sn; and verifies that the ticket has not been spent:

---

[4]A communication channel is *resilient* if a message inserted into such a channel will eventually be delivered.

a. if $\nexists \psi_{\gamma_i,0}$ linked to **PASS*** in the database:
  P_i follows the next steps:

  i. generates a challenge $c \xleftarrow{R} \mathbb{Z}_q$;
  ii. assigns *Challenge* $= (\mathsf{PASS.Sn}, c, \tau_1)$;
  iii. *Challenge*$^* = (Challenge, \mathsf{sk}_{\mathscr{P}i}(hash(Challenge)))$ and sends this signature to $\mathscr{U}$;
  iv. asynchronously, for optimization, pre-computes $\mathsf{H}_{\mathscr{U}}^c \pmod p$; $\mathscr{U}$ computes:

    i. computes $w_3 = a_3 + c \cdot \mathsf{RU} \pmod q$;
    ii. encrypts and signs w_3 and, then, sends it to P_i: $\mathsf{sk}_{\mathscr{U}}(\mathsf{pk}_{\mathscr{P}i} (\mathsf{PASS.Sn}, w_3, \tau_1))$;
      P_i follows the next steps:

      i. computes $\alpha^{w_3} \pmod p$;
      ii. verifies $\alpha^{w_3} \overset{?}{=} A_3 \cdot \mathsf{H}_{\mathscr{U}}^c \pmod p$;
      → **if verification fails**:
      iii. assigns $\mathsf{V}_{\mathsf{fail}} = (\mathsf{PASS.Sn}, \mathsf{flag}_{01}, \tau_1)$.
      iv. signs $\mathsf{V}_{\mathsf{fail}}^* = (\mathsf{V}_{\mathsf{fail}}, \mathsf{sk}_{\mathscr{P}i}(hash(\mathsf{V}_{\mathsf{fail}})))$ and sends $\mathsf{m}_2 = \mathsf{V}_{\mathsf{fail}}^*$ to $\mathscr{U}$
      → **else**:

  iii. computes $\mathsf{A}_{\mathscr{P}} = PRNG(\mathsf{h_K}) \oplus \mathsf{RI}_{\gamma_i}$, where $PRNG(\mathsf{h_K})$ is a secure pseudorandom number generator and, $\mathsf{h_K} = hash(\mathsf{K})$ is the seed. Note that K and RI_{γ_i} are obtained from $\delta_{\mathscr{T},\mathscr{P}i}$;
  iv. encrypts $\mathsf{A}_{\mathscr{P}}$ with the public key of the TTP $\mathscr{T}$: $\mathsf{pk}_{\mathscr{T}}(\mathsf{A}_{\mathscr{P}})$;
  v. assigns $\mathsf{V}_{\mathsf{succ}} = (\gamma_i, \mathsf{PASS.Sn}, \mathsf{flag}_{10}, \tau_1, \mathsf{pk}_{\mathscr{T}}(\mathsf{A}_{\mathscr{P}}))$, ($\tau_1$ is the verification timestamp). The signature is noted: $\mathsf{V}_{\mathsf{succ}}^* = (\mathsf{V}_{\mathsf{succ}}\mathsf{sk}_{\mathscr{P}i}, (hash(\mathsf{V}_{\mathsf{succ}})))$;
  vi. sends $\mathsf{m}_2 = \mathsf{V}_{\mathsf{succ}}^*$ to $\mathscr{U}$;

b. if $\exists \psi_{\gamma_i,0}$ linked to **PASS*** in the database:

  i. assigns $\mathsf{V}_{\mathsf{fail}} = (\mathsf{PASS.Sn}, \psi_{\gamma_i,0}, \mathsf{flag}_{02}, \tau_1, \gamma_i)$. The signature is noted: $\mathsf{V}_{\mathsf{fail}}^* = (\mathsf{V}_{\mathsf{fail}}, \mathsf{sk}_{\mathscr{P}i}(hash(\mathsf{V}_{\mathsf{fail}})))$;
  ii. sends $\mathsf{m}_2 = \mathsf{V}_{\mathsf{fail}}^*$ to $\mathscr{U}$; indicating that this non reusable ticket of the PASS had been already used.

showProof $\mathscr{U}$ executes:

1. verifies P_i's signature;
→ **if $\mathsf{V}_{\mathsf{fail}}^*$ is received or $\mathsf{V}_{\mathsf{succ}}^*$ is not correct**:
2. $\mathscr{U}$ checks the appropriate **flag** to know the details of the error. If he does not agree then he can rise a CLAIM by contacting with $\mathscr{T}$;
→ **else**:

3. calculates $A_{\mathcal{U}} = PRNG(K) \oplus \psi_{\gamma_i,0}$, using the shared value K as seed;
4. compose the message $m_3 = (PASS.Sn, A_{\mathcal{U}})$;
5. signs and sends it to P_i: $m_3^* = (m_3, sk_{\mathcal{P}i}(hash(m_3)))$;

verifyProof P_i follows the next steps:

1. obtains $PASS.Sn$, and computes $\psi_{\gamma_i,0} = A_{\mathcal{U}} \oplus PRNG(K)$;
2. verifies $\psi_{\gamma_i,1} \overset{?}{=} hash(\psi_{\gamma_i,0})$;
3. generates τ_2 and verifies it using the PASS expiry date (PASS.PURdate, PASS.EXPdate) and the timestamp τ_1, the value of ACTdate of the element ACT and LIMdate, being LIMdate = ACTdate + Lifetime;

→ **if any verification fails**:
4. assigns $V_{fail} = (\gamma_i, PASS.Sn, flag_{03}, \tau_2, \psi_{\gamma_i,0})$.
5. signs $V_{fail}^* = (V_{fail}, sk_{\mathcal{P}i}(hash(V_{fail})))$, sends $m_2 = V_{fail}^*$ to $\mathcal{U}$

→ **else**:
4. signs $A_{\mathcal{P}}$ 1approving then the validation with timestamp τ_2: $R_{\gamma_i} = (A_{\mathcal{P}}, PASS.Sn, \tau_2)$, and $R_{\gamma_i}^* = (R_{\gamma_i}, sk_{\mathcal{P}}(hash(R_{\gamma_i})))$;
5. stores in its *SpentCityPASSES* database: $(PASS^*, \psi_{\gamma_i,0})$ and sends $m_4 = R_{\gamma_i}^*$ to $\mathcal{U}$;

getValidationConfirmation $\mathcal{U}$ follows the next steps:

→ **if V_{fail}^* is received**:
1. $\mathcal{U}$ checks $flag_{03}$ to know the details of the error. If he does not agree then he can rise a CLAIM by contacting with the $\mathcal{T}$;

→ **else**:
1. checks the signature of $R_{\gamma_i}^*$;
2. computes $RI_{\gamma_i} = A_{\mathcal{P}} \oplus PRNG(h_K)$;
3. verifies $h_{RI_{\gamma_i}} \overset{?}{=} hash(RI_{\gamma_i})$;

→ **if any verification fails**:
4. $\mathcal{U}$ collects all evidence and he can rise a CLAIM by contacting with the $\mathcal{T}$, he can argue that there is an error in getting the authorisation to use the service (*Authorisation Failure*);

→ **else**:
4. stores in his *TicketsPASS* database $(R_{\gamma_i}^*, RI_{\gamma_i})$ together with $PASS^*$.

Verification of a *m*-times reusable service λ_i, the verification of the corresponding ticket is as follows:

showPASS $\mathcal{U}$ computes:

1. generates a random value $a_3, \overset{R}{\leftarrow} \mathbb{Z}_q$ to be used in a Schnorr proof;
2. computes $A_3 = \alpha^{a_3} \pmod{p}$;

3. **Case 1-First time use of a m-times reusable service**: $\mathcal{U}$ computes $\psi_{\lambda_i, m-1} = hash^{m-1}(\psi_{\lambda_i, 0})$. Then, $\mathcal{U}$ creates a counter $k_{\lambda_i} = m - 1$ to keep track of the number of times that the ticket can be used. The counter is stored in his *CityPASS* database together with the rest of the information associated to this PASS ticket.[5]

3. **Case 2-Subsequent use of a m-times reusable service**: $\mathcal{U}$ retrieves k_{λ_i} from her *TicketsPASS* database and computes $\psi_{\lambda_i, (k_{\lambda_i}-1)} = hash^{(k_{\lambda_i}-1)}(\psi_{\lambda_i, 0})$.

4. composes the information ticket message $m_1 = (\mathsf{PASS}^*, \mathsf{ACT}^*, A_3, (k_{\lambda_i}))$;

5. signs and sends it to P_i: $m_1^* = (m_1, \mathsf{sk}_{\mathscr{P}i}(hash(m_1)))$;

verifyPASS P_i executes:

1. verifies the PASS signature, PASS.PURdate, and PASS.EXPdate;
2. verifies ACT^*, and PASS.ACTdate;
→ **if any verification fails**:
3. assigns $\mathsf{V_{fail}} = (\mathsf{PASS.Sn}, \mathsf{flag}_{00}, \tau_1)$.
4. signs $\mathsf{V_{fail}^*} = (\mathsf{V_{fail}}, \mathsf{sk}_{\mathscr{P}i}(hash(\mathsf{V_{fail}})))$ and sends $m_2 = \mathsf{V_{fail}^*}$ to $\mathcal{U}$
→ **else**:

3. P_i looks for the PASS^* in its *SpentCityPASSES* database using PASS.Sn; and verifies that the ticket has not been already spent m-times:

    a. if $(\nexists \psi_{\lambda, k_{\lambda_i}}$ linked to $\mathsf{PASS}^*)$ or $[(\exists \psi_{\lambda, k_{\lambda_i}})$ and $(k_{\lambda_i} \geq 1)$ and $(k_{\lambda_i}$ stored in the P_i database has the same value than the one sent by $\mathcal{U})]$: P_i follows the next steps:

        i. generates a challenge $\mathsf{c} \xleftarrow{R} \mathbb{Z}_q$;
        ii. assigns *Challenge* $= (\mathsf{PASS.Sn}, \mathsf{c}, \tau_1)$;
        iii. *Challenge*$^* = (Challenge, \mathsf{sk}_{\mathscr{P}i}(hash(Challenge)))$ and sends this signature to $\mathcal{U}$;
        iv. asynchronously, for optimization, pre-computes $\mathsf{H}_{\mathcal{U}}^{\mathsf{c}} (mod\, p)$; $\mathcal{U}$ computes:

        i. computes $\mathsf{w}_3 = a_3 + \mathsf{c} \cdot \mathsf{RU} \ (mod\, q)$;
        ii. encrypts and signs w_3 and, then, sends it to P_i: $\mathsf{sk}_{\mathcal{U}}(\mathsf{pk}_{\mathscr{P}i}(\mathsf{PASS.Sn}, w_3, \tau_1))$; P_i follows the next steps:

        i. computes $\alpha^{\mathsf{w}_3} (mod\, p)$;
        ii. verifies $\alpha^{\mathsf{w}_3} \stackrel{?}{=} A_3 \cdot \mathsf{H}_{\mathcal{U}}^{\mathsf{c}} (mod\, p)$;
        → **if verification fails**:
        iii. assigns $\mathsf{V_{fail}} = (\mathsf{PASS.Sn}, \mathsf{flag}_{01}, \tau_1)$.
        iv. signs $\mathsf{V_{fail}^*} = (\mathsf{V_{fail}}, \mathsf{sk}_{\mathscr{P}i}(hash(\mathsf{V_{fail}})))$ and sends $m_2 = \mathsf{V_{fail}^*}$ to $\mathcal{U}$
        → **else**:

---

[5]Note that $\psi_{\lambda_i, 0}$ was stored by $\mathcal{U}$ in her *TicketsPASS* database at the PASS purchase phase.

       iii. computes $A_{\mathscr{P}} = PRNG(h_K) \oplus RI_{\lambda_i}$, where $PRNG(h_K)$ is a secure pseudorandom number generator and, $h_K = hash(K)$ is the seed. Note that K and RI_{λ_i} are obtained from $\delta_{\mathscr{T},\mathscr{P}_i}$;

       iv. encrypts $A_{\mathscr{P}}$ with the public key of the TTP $\mathscr{T}$: $pk_{\mathscr{T}}(A_{\mathscr{P}})$;

       v. assigns $V_{succ} = (\lambda_i, PASS.Sn, flag_{10}, \tau_1, pk_{\mathscr{T}}(A_{\mathscr{P}}), k_{\lambda_i}, \psi_{(\lambda,k_{\lambda_i})})$. The signature is noted: $V_{succ}^* = (V_{succ}, sk_{\mathscr{P}i}(hash(V_{succ})))$;

       vi. sends $m_2 = V_{succ}^*$ to $\mathscr{U}$;

   b. else[6]:

       i. assigns $V_{fail} = (\lambda_i, PASS.Sn, flag_{04}, \tau_1, k_{\lambda_i}, \psi_{(\lambda,k_{\lambda_i})})$. The signature is noted: $V_{fail}^* = (V_{fail}, sk_{\mathscr{P}}(hash(V_{fail})))$;

       ii. sends $m_2 = V_{fail}^*$ to $\mathscr{U}$;

showProof $\mathscr{U}$ executes:

1. verifies P_i's signature;

→ **if V_{fail}^* is received or V_{succ}^* is not correct**:

2. $\mathscr{U}$ checks the appropriate flag to know the details the error. If he does not agree then he can rise a CLAIM by contacting with $\mathscr{T}$;

→ **else**:

2. calculates $A_{\mathscr{U}} = PRNG(K) \oplus \psi_{\lambda,(k_{\lambda_i}-1)}$, using the shared value K as seed;

3. compose the message $m_3 = (PASS.Sn, A_{\mathscr{U}})$;

4. signs and sends it to P_i: $m_3^* = (m_3, sk_{\mathscr{P}i}(hash(m_3)))$;

verifyProof P_i follows the next steps:

1. obtains $PASS.Sn$, and computes $\psi_{\lambda,(k_{\lambda_i}-1)} = A_{\mathscr{U}} \oplus PRNG(K)$;

2. checks $\psi_{\lambda,(k_{\lambda_i})} \stackrel{?}{=} hash(\psi_{\lambda,(k_{\lambda_i}-1)})$,

3. generates τ_2 and verifies it using the PASS expiry date ($PASS.PURdate, PASS.EXPdate$) and the timestamp τ_1;

→ **if any verification fails**:

4. assigns $V_{fail} = (\lambda_i, PASS.Sn, flag_{05}, \tau_2, \psi_{\lambda,(k_{\lambda_i}-1)}, (k_{\lambda_i} - 1))$.

5. signs $V_{fail}^* = (V_{fail}, sk_{\mathscr{P}i}(hash(V_{fail})))$, sends $m_2 = V_{fail}^*$ to $\mathscr{U}$

→ **else**:

4. signs $A_{\mathscr{P}}$ approving then the validation with timestamp τ_2: $R_{\lambda_i} = (A_{\mathscr{P}}, PASS.Sn, \tau_2)$, and $R_{\lambda_i}^* = (R_{\lambda_i}, sk_{\mathscr{P}}(hash(R_{\lambda_i})))$;

5. stores in the *SpentCityPASSES* database: $(PASS^*, \psi_{\lambda,(k_{\lambda_i}-1)})$;

---

[6]An error has been detected, and thus the ticket is not valid. The reason of the error can be:

- k_{λ_i} sent by $\mathscr{U}$ is greater than the one stored by P_i
- $\psi_{(\lambda,k_{\lambda_i}-1)}$ is not correct
- the *PASS* for the λ_i service is over ($k_{\lambda_i} = 0$).

6. updates the value of $k_{\lambda_i} = (k_{\lambda_i} - 1)$ and stored it with the ticket information;
7. sends $m_4 = R^*_{\lambda_i}$ to $\mathcal{U}$;

getValidationConfirmation $\mathcal{U}$ follows the next steps:

$\rightarrow$ **if V^*_{fail} is received**:
1. $\mathcal{U}$ checks flag to know the details the error. If he does not agree then he can rise a CLAIM by contacting with the $\mathcal{T}$;
$\rightarrow$ **else**:
1. checks the signature of $R^*_{\lambda_i}$;
2. computes $RI_{\lambda_i} = A_{\mathscr{P}} \oplus PRNG(h_K)$;
3. verifies $h_{RI_{\lambda_i}} \overset{?}{=} hash(RI_{\lambda_i})$;
$\rightarrow$ **if any verification fails**:
4. $\mathcal{U}$ collects evidences and he can rise a CLAIM contacting with $\mathcal{T}$;
$\rightarrow$ **else**:
4. stores in her *TicketsPASS* database $(R^*_{\lambda_i}, RI_{\lambda_i})$ together with PASS* and $k_{\lambda_i} = (k_{\lambda_i} - 1)$.

    Verification of an infinitely-reusable service ξ_i, service provider P_i and user $\mathcal{U}$ follow a verification phase similar to the verification of a non-reusable service. In this case, however, if the user presents a valid R^*_ξ and the current date is lower than LIMdate, the service can be used again.

4 Claims

The *PASS Verification* protocol has been designed as a fair exchange protocol (a valid e-ticket is given in exchange for the permission to use the service). We introduce an offline TTP [22] between the user and the provider of the service in order to guarantee the fairness of the exchange. Fairness guarantees that either each party ($\mathcal{U}$ or P_i) receives the item it expects or neither party receives any additional information about the other's item. In case of exception (i.e. an unfair situation has come up and the exchange does not ends successfully for some party), any actor can arise a claim in order to restore the fairness of the exchange and preserve the security of the system.

    In the *PASS Verification* there are some cases where users can receive a wrong token from the providers that can break the fairness of the system. In theses cases, users can contact the TTP to start a new claim in order to preserve the fairness of the exchange. If any claim arises, then the $\mathcal{T}$ acts as an unbiased arbiter between parties. $\mathcal{T}$ will evaluate the presented evidence and then it will side by one party depending on the result of the verifiable evidence. Thus, the involved parties have to fix the situation according to the $\mathcal{T}$ decision.

4.1 Non-reusable CLAIMS

Next, we are going to summarize the claims that can arise in the *PASS Verification* protocol for non-reusable services. In all cases, at the end of the verifications, $\mathcal{T}$ will publish its resolution according to them.

4.1.1 flag$_{00}$ CLAIM

$\mathcal{U}$ has received $\mathsf{V_{fail}}$ from P_i indicating whether any parameter of the PASS sent by $\mathcal{U}$ has not the proper form or its signature are not valid. If $\mathcal{U}$ does not agree with P_i's, then he can submit the following items to $\mathcal{T}$:

- $\mathsf{V_{fail}} = (\mathsf{PASS.Sn}, \mathsf{flag}_{00}, \tau_1); \| \mathsf{V_{fail}^*} = (\mathsf{V_{fail}}, \mathsf{sk}_{\mathscr{P}i}(hash(\mathsf{V_{fail}})))$
- $\mathsf{PASS^*} = (\mathsf{PASS}, \mathsf{Sign}_{\mathscr{I}}(\mathsf{PASS}))$;

$\mathcal{T}$ can verify that the items are linked by the $\mathsf{PASS.Sn}$, next it can check the signature and the correctness of each message (including the τ_1 and the dates inside the PASS).

4.1.2 flag$_{01}$ CLAIM

$\mathcal{U}$ has received $\mathsf{V_{fail}}$ from P_i indicating that $\mathcal{U}$ has not been able to demonstrate the ownership of the PASS (there was a problem during the ZKP protocol run). If $\mathcal{U}$ does not agree with the P_i's message, then he can submit the following items to the $\mathcal{T}$:

- $\mathsf{pk}_{\mathscr{P}i}(\mathsf{PASS.Sn}, \mathsf{w}_3, \tau_1)$;
- $\mathsf{V_{fail}} = (\mathsf{PASS.Sn}, \mathsf{flag}_{01}, \tau_1); \| \mathsf{V_{fail}^*} = (\mathsf{V_{fail}}, \mathsf{sk}_{\mathscr{P}i}(hash(\mathsf{V_{fail}})))$;
- *Challenge**; m_1; $\mathsf{PASS^*}$;

$\mathcal{T}$ can verify that the items are linked by the $\mathsf{PASS.Sn}$, next it can check the signature and the correctness of each message. In addition, $\mathcal{T}$ asks P_i to decrypt $\mathsf{pk}_{\mathscr{P}i}(\mathsf{PASS.Sn}, \mathsf{w}_3, \tau_1)$. Then it will check whether the ZKP's response is correct or not.

4.1.3 flag_{10} CLAIM

$\mathscr{U}$ has received V_{succ} from P_i indicating that the ticket is valid and has not been spent yet. However, if $\mathscr{U}$ detects that some of the parameters of this message are not correct, then he can submit the following items to the $\mathscr{T}$:

- $V_{\text{succ}} = (\gamma_i, \text{PASS.Sn}, \text{flag}_{10}, \tau_1, \text{pk}_{\mathscr{T}}(A_{\mathscr{P}})) \parallel V_{\text{succ}}^* = (V_{\text{succ}}, \text{sk}_{\mathscr{P}i}(hash(V_{\text{succ}})));$
- $m_1; \text{PASS}^*;$

$\mathscr{T}$ can verify that the items are linked by the PASS.Sn, next it can check the signature and the correctness of each message.

4.1.4 flag_{02} CLAIM

$\mathscr{U}$ has received V_{fail} from P_i indicating that the ticket has been spent, thus it is not valid anymore. But, if $\mathscr{U}$ does not agree with the P_i's message, then he can submit the following items to the $\mathscr{T}$:

- $V_{\text{fail}} = (\text{PASS.Sn}, \psi_{\gamma_i,0}, \text{flag}_{02}, \tau_1, \gamma_i) \parallel V_{\text{fail}}^* = (V_{\text{fail}}, \text{sk}_{\mathscr{P}i}(hash(V_{\text{fail}})));$
- $m_1; \text{PASS}^*;$

$\mathscr{T}$ can verify whether the evidence $\psi_{\gamma_i,0}$ unveiled P_i is correct or not. Thus, it is able to check if the PASS has been already spent.

4.1.5 flag_{03} CLAIM

$\mathscr{U}$ has received V_{fail} from P_i indicating two possible errors: $\psi_{\gamma_i,0}$ does not match with $\psi_{\gamma_i,1}$ ($\psi_{\gamma_i,1} = hash(\psi_{\gamma_i,0})$), or τ_2 does not match the expiry date of the PASS. But, if $\mathscr{U}$ does not agree with the error message, then he can submit the following items to the $\mathscr{T}$:

- $V_{\text{fail}} = (\gamma_i, \text{PASS.Sn}, \text{flag}_{03}, \tau_2, \psi_{\gamma_i,0}) \parallel V_{\text{fail}}^* = (V_{\text{fail}}, \text{sk}_{\mathscr{P}i}(hash(V_{\text{fail}}))) \parallel m_2 = V_{\text{fail}}^*$
- $m_3 = (\text{PASS.Sn}, A_{\mathscr{U}})$

$\mathscr{T}$ can verify the evidence $\psi_{\gamma_i,1} \overset{?}{=} hash(\psi_{\gamma_i,0})$ because it has access to K (see Sect. 3.1.3). Also, it can verify whether τ_2 is correct or not according to the expiry date of the PASS.

4.1.6 Authorization Failure CLAIM

$\mathscr{U}$ has received a wrong R_{γ_i}. Therefore he is not able to use the service. In order to solve the situation, he can submit the following items to $\mathscr{T}$:

- $R_{\gamma_i} = (A_{\mathscr{P}}, PASS.Sn, \tau_2) || R_{\gamma_i}^* = (R_{\gamma_i}, sk_{\mathscr{P}}(hash(R_{\gamma_i}))) || m_4 = R_{\gamma_i}^*$
- $PASS^* = (PASS, Sign_{\mathscr{I}}(PASS))$;

$\mathscr{T}$ can decrypt $A_{\mathscr{P}}$ evidence, thus can check whether the decrypted value RI_{γ_i} matchs $hash(RI_{\gamma_i})$ stored in the PASS.

4.2 Reusable CLAIMS

This section summarizes the claims that can arise in the *PASS Verification* protocol for *m*-times reusable services.

4.2.1 flag$_{00}$ CLAIM

$\mathscr{U}$ has received V_{fail} from P_i indicating if any parameter of the PASS sent by $\mathscr{U}$ has not the proper form or its signature is not valid. If $\mathscr{U}$ does not agree with P_i's message, then he can submit the following items to $\mathscr{T}$:

- $V_{fail} = (PASS.Sn, flag_{00}, \tau_1); || V_{fail}^* = (V_{fail}, sk_{\mathscr{P}i}(hash(V_{fail})))$
- $PASS^* = (PASS, Sign_{\mathscr{I}}(PASS))$;

$\mathscr{T}$ can verify that the items are linked by the PASS.Sn, next it can check the signature and the correctness of each message (e.g. the τ_1 is past the final expiry inside the PASS).

4.2.2 flag$_{01}$ CLAIM

$\mathscr{U}$ has received V_{fail} from P_i indicating that $\mathscr{U}$ has not been able to demonstrate the ownership of the PASS (there was a problem during the ZKP protocol run). If $\mathscr{U}$ does not agree with the P_i's message, then he can submit the following items to the $\mathscr{T}$:

- $pk_{\mathscr{P}i}(PASS.Sn, w_3, \tau_1)$;
- $V_{fail} = (PASS.Sn, flag_{01}, \tau_1); || V_{fail}^* = (V_{fail}, sk_{\mathscr{P}i}(hash(V_{fail})))$;
- *Challenge**; m_1; PASS*;

$\mathscr{T}$ can verify that the items are linked by the PASS.Sn, next it can check the signature and the correctness of each message. In addition, $\mathscr{T}$ asks P_i to decrypt $\mathsf{pk}_{\mathscr{P}_i}(\mathsf{PASS.Sn}, w_3, \tau_1)$. Then, it will check if the ZKP's response is correct.

4.2.3 flag$_{10}$ CLAIM

$\mathscr{U}$ has received $\mathsf{V}_{\mathsf{succ}}$ from P_i indicating that the ZKP has finished successfully (i.e. $\mathscr{U}$ has proven the ownership of the PASS) and, also, P_i agrees with counter k_{λ_i}. So, the service can still be used by $\mathscr{U}$ using the *PASS*. However, if $\mathscr{U}$ detects that some of the parameters of this message are not correct, then he can submit the following items to the $\mathscr{T}$:

- $\mathsf{V}_{\mathsf{succ}} = (\lambda_i, \mathsf{PASS.Sn}, \mathsf{flag}_{10}, \tau_1, \mathsf{pk}_{\mathscr{T}}(A_{\mathscr{P}}), k_{\lambda_i}, \psi_{(\lambda, k_{\lambda_i})}) || \mathsf{V}_{\mathsf{fail}}^*$
  $= (\mathsf{V}_{\mathsf{fail}}, \mathsf{sk}_{\mathscr{P}_i}(hash(\mathsf{V}_{\mathsf{fail}})));$

- m_1; PASS*;

$\mathscr{T}$ can verify that the items are linked by the PASS.Sn, next it can check the signature and the correctness of each message.

4.2.4 flag$_{04}$ CLAIM

$\mathscr{U}$ has received $\mathsf{V}_{\mathsf{fail}}$ from P_i indicating whether any parameter of the PASS sent by $\mathscr{U}$ are not correct[6]. If $\mathscr{U}$ does not agree with the P_i's message, then he can submit the following items to the $\mathscr{T}$:

- $\mathsf{V}_{\mathsf{fail}} = (\lambda_i, \mathsf{PASS.Sn}, \mathsf{flag}_{04}, \tau_1, k_{\lambda_i}, \psi_{(\lambda, k_{\lambda_i})}) || \mathsf{V}_{\mathsf{fail}}^* = (\mathsf{V}_{\mathsf{fail}}, \mathsf{sk}_{\mathscr{P}_i}(hash(\mathsf{V}_{\mathsf{fail}})));$

- m_1; PASS*;

$\mathscr{T}$ can verify that the items are linked by the PASS.Sn, next it can check the signature and the correctness of each message (e.g. whether $\psi_{(\lambda, k_{\lambda_i})}$ proposed by P_i in $\mathsf{V}_{\mathsf{fail}}$ is equal to $hash^{(k_{\lambda_i})}(\psi_{\lambda_i,0})$ or not).

4.2.5 flag$_{05}$ CLAIM

$\mathscr{U}$ has received $\mathsf{V}_{\mathsf{fail}}$ from P_i indicating two possible errors: $\psi_{\lambda,(k_{\lambda_i}-1)}$ does not match $(\psi_{\lambda,(k_{\lambda_i})} \stackrel{?}{=} hash(\psi_{\lambda_i,(k_{\lambda_i}-1)}))$, or τ_2 does not match the expiry date of the PASS. But, if $\mathscr{U}$ does not agree with the error message, then he can submit the following items to the $\mathscr{T}$:

- $V_{fail} = (\lambda_i, PASS.Sn, flag_{05}, \tau_2, \psi_{\lambda,(k_{\lambda_i}-1)}, (k_{\lambda_i}-1))||V^*_{fail}$
  $= (V_{fail}, sk_{\mathcal{P}i}(hash(V_{fail})))||m_2 = V^*_{fail};$
- $m_3 = (PASS.Sn, A_{\mathcal{U}})$

$\mathcal{T}$ can verify $\psi_{\lambda,(k_{\lambda_i}-1)}$ evidence, because it has access to K (see Sect. 3.1.3). Also, it can verifies whether τ_2 is correct or nor according to the expiry date of the PASS.

4.2.6   Authorization Failure CLAIM

$\mathcal{U}$ has received a wrong R_{λ_i} and, therefore he is not able to reach use the service. In order to solve the situation, he can submit the following items to the $\mathcal{T}$:

- $R_{\lambda_i} = (A_{\mathcal{P}}, PASS.Sn, \tau_2)||R^*_{\lambda_i} = (R_{\lambda_i}, sk_{\mathcal{P}}(hash(R_{\lambda_i})))||m_4 = R^*_{\lambda_i}$
- $PASS^* = (PASS, Sign_{\mathcal{I}}(PASS));$

$\mathcal{T}$ can decrypt $A_{\mathcal{P}}$ evidence, thus can check whether the decrypted value RI_{λ_i} matchs $hash(RI_{\lambda_i})$ stored in the PASS.

5   System Deployment and Analysis

Previous sections include the description of the protocol used in our proposal. However, in order to have a functional solution we have to proof the security of the protocol through a security and privacy analysis and its viability through a performance evaluation after the implementation of the protocol.

The analysis of the requirements of the system can be performed from both the functionality and the security-privacy point of view. The functional requirements, such as reduced size, flexibility, easyness of use, efficiency and so on, will be evaluated after the implementation of the system.

For the security-privacy point of view we have written a formal security analysis based in the list of requirements for electronic tickets, both listed in Sect. 2.1 and included in the survey [4] but due to space constrains we cannot include it in the paper. We have been able to prove that the mCITYPASS system complies with the requirements of authenticity, non-repudiation, integrity, reusability, unforgeability, fairness, non-overspending, unsplittability and exculpability. For the peculiarities of a CITYPASS system transferability is not contemplated since the PASS is intended for its usage by a single user. Exculpability and fairness can be assured thanks to the claims described in Sect. 4. In terms of privacy, the user is protected due to the revocable anonymity of the system. Only the identity of fraudulent users will be disclosed.

Currently we are working on the implementation of the protocol and we are obtaining some preliminar results about the execution performance.

6 Conclusions and Further Works

Several proposals have been presented describing electronic ticketing systems [5, 8, 11, 18, 23, 24]. These proposals are general purpose systems or applied to transport services. In this paper we have presented an electronic ticketing system intended to be used for touristic services. Almost all big cities have inefficient CITYPASS systems usually implemented on smartcards. Our proposal is the first one that can be implemented on portable devices, such as smartphones, and is flexible enough to include reusable and non-reusable services in the same PASS. The system has been performed taking into account all the security and privacy requirements described for electronic tickets, including the challenging ones (exculpability, reusability and unsplittability) resulting in a very secure and powerful system. Moreover, the dispute resolution protocol assures that all the parts are protected against other part's attacks. Finally, the system allows the user to use the system anonymously, so the privacy of the system is assured. As a future work, we want to finish the implementation of the system and obtain experimental results in order to verify the usability and performance of the system. Moreover, the security and performance analysis will be presented in a formal way.

Disclaimer and Acknowledgements This work was partially supported by the Spanish Government under AccessTur TIN2014-54945-R AEI/FEDER UE and Red de excelencia Consolider ARES TIN2015-70054-REDC projects.

References

1. European Parliamentary Research Service: Integrated urban e-ticketing for public transport and touristic sites (2014)
2. City Cards. We love city cards, Date of access, June. http://welovecitycards.com (2017)
3. Inc. City Pass, Date of access: June, 2017. http://citypass.com
4. Mut-Puigserver, M., Magdalena Payeras-Capellà, M., Ferrer-Gomila, J.-L., Vives-Guasch, A, Castellà-Roca, J.: A survey of electronic ticketing applied to transport. Comput. Secur. 31(8), 925–939 (2012)
5. Vives-Guasch, A., Magdalena Payeras-Capellà, M., Mut-Puigserver, M., Castellà-Roca, J., Ferrer-Gomila, J.-L.: A secure e-ticketing scheme for mobile devices with near field communication (NFC) that includes exculpability and reusability. IEICE Trans. 95-D(1), 78–93 (2012)
6. European Union. Cleaner and better transport in cities, civitas 2020, Date of access: June, 2017. http://www.civitas.eu/content/integrated-e-ticketing-system

7. Mallat, N., Rossi, M., Tuunainen, V.K., Öörni, A.: An empirical investigation of mobile ticketing service adoption in public transportation. Pers. Ubiquit. Comput. **12**(1), 57–65 (2008)
8. Bohm, A., Murtz, B., Sommer, C., Wermuth, M.: Location-based ticketing in public transport. In Proceedings of 8th International IEEE Conference on Intelligent Transportation Systems, vol. 12, pp. 12:837–840 (2005)
9. Angeloni, S.: A tourist kit made in Italy: an intelligent system for implementing new generation destination cards. Tourism Manage. **52**, 187–209 (2016)
10. Chow, J.Y.J.: Policy analysis of third party electronic coupons for public transit fares. Transp. Res **66**(1), 238–250 (2014)
11. Kos-Labedowicz, J.: Integrated E-ticketing System—Possibilities of Introduction in EU, pp. 376–385. Springer, Berlin (2014)
12. Basili, A., Liguori, W., Palumbo, F.: Nfc smart tourist card: combining mobile and contactless technologies towards a smart tourist experience. In IEEE 23rd International WETICE Conference, pp. 249–254, June (2014) doi:10.1109/WETICE.2014.61
13. Bang, S.W., Park, K.J., Kim, W.S., Park, G.D., Im, D.H.: Design and implementation of NFC-based mobile coupon for small traders and enterprisers. In 2013 International Conference on IT Convergence and Security (ICITCS), pp. 18–20 (2013). doi:10.1109/ICITCS.2013.6717820
14. Dominikus, S., Aigner, M.: mcoupons: an application for near field communication (NFC). In: 21st International Conference on Advanced Information Networking and Applications Workshops, 2007, AINAW '07, vol. 2, pp. 421–428, May (2007)
15. Hsiang, H.C.: A secure and efficient authentication scheme for m-coupon systems. In: 2014 8th International Conference on Future Generation Communication and Networking, pp 17–20, Dec (2014)
16. Liu, X., Xu, Q.L.: Practical compact multi-coupon systems. In 2009 IEEE International Conference on Intelligent Computing and Intelligent Systems, vol. 3, pp. 211–216, Nov (2009)
17. Chen, L., Escalante B., A.N., Löhr, H., Manulis, M., Sadeghi, A.-R.: A Privacy-Protecting Multi-Coupon Scheme with Stronger Protection Against Splitting, pp. 29–44. Springer, Berlin (2007)
18. Armknecht, F., Escalante B., A.N., Löhr, H., Manulis, M., Sadeghi, A.-R.: Secure Multi-Coupons for Federated Environments: Privacy-Preserving and Customer-Friendly, pp. 29–44. Springer, Berlin (2008)
19. Stallings, W.: Cryptography and Network Security: Principles and Practice, 7th ed. Pearson Education Ltd (2017). ISBN: 13:978-0134444284
20. Dingledine, R., Mathewson, N., Syverson, P.: Tor: the second-generation onion router. In 13th USENIX Security Symposium (2004)
21. Schnorr, C.-P.: Efficient signature generation by smart cards. J. Cryptology **4**(3), 161–174 (1991)
22. Kremer, S., Markowitch, O., Zhou, J.: An intensive survey of fair nonrepudiation protocols. Comput. Commun. **25**, 1606–1621 (2002)
23. Chen, Y.-Y., Chen, C.-L., Jan, J.-K.: A mobile ticket system based on personal trusted device. Wirel. Pers. Commun. Int. J. **40**(4), 569–578 (2007)
24. Quercia, D., Hailes, S.: Motet: Mobile transactions using electronic tickets. In: 1st International Conference on Security and Privacy for Emerging Areas in Communications Networks, Proceedings, vol. 24, pp. 374–383, Athens, Greece, Sept (2005)

Heuristic-Based Usability Evaluation Tool for Android Applications

Kwandee Phetcharakarn and Twittie Senivongse

Abstract Heuristic evaluation is a popular method for evaluating usability of user interface due to its simplicity and cost efficiency compared to other methods. Evaluation is done by evaluators by inspecting user interface on each screen of an application against usability design principles or heuristics. The evaluation depends on judgment and experience of the evaluators whether they can spot the problems, and there are many heuristics to check against. This paper attempts to assist the evaluators by automating the evaluation task against a number of design heuristics. The paper presents a development of a usability evaluation tool for Android applications by inspecting source code and reporting locations in the code where violations of heuristics are found. Although only small part of the heuristics can be checked automatically, an experiment shows that the tool can help save evaluation time and can discover usability problems that are missed by evaluators.

Keywords Heuristic evaluation · Usability · Android

1 Introduction

Usability of applications refers to how easy the user interfaces of the applications are to use. By easy, the term embraces five quality components of the interface design [1]: (1) Learnability involves how easy the user interface is for the users to accomplish tasks the first time they use, (2) Efficiency involves how quickly the users can perform tasks once they learn the design. (3) Memorability involves how easily the users can reestablish proficiency after not using the interface for a period of time. (4) Errors involves how well the design can prevent the users from making

K. Phetcharakarn · T. Senivongse (✉)
Department of Computer Engineering, Faculty of Engineering, Chulalongkorn University,
10330 Bangkok, Thailand
e-mail: twittie.s@chula.ac.th

K. Phetcharakarn
e-mail: kwandee.p@student.chula.ac.th

© Springer International Publishing AG 2018   161
R. Lee (ed.), *Applied Computing & Information Technology*,
Studies in Computational Intelligence 727, DOI 10.1007/978-3-319-64051-8_10

many errors or severe errors and how easily the users can recover from any errors that occur. and (5) Satisfaction involves how pleasant it is to use the design. Like other kinds of applications, the design of mobile user interfaces has to take into account these quality components. Unlike other kinds of applications, the design has to consider unique characteristics of mobile devices, i.e. limitation on screen size, resolution, and resources, touch and gesture-based interaction, and touchless interaction such as speech.

To evaluate how usable the user interface design is, a popular method called heuristic evaluation [1] can be used. It is an inspection method in which evaluators, preferably usability experts, compare the user interface design of a software product with a list of design guidelines (or heuristics) and identify where the design does not follow the guidelines. Different levels of guidelines are available [1]. General guidelines are applicable to all user interfaces and usually are broad design principles. More specific guidelines are available for the specific category of the system being developed or even for the individual product. Many times, when trained usability experts are not available, developers with some experience in user interface design can do the review themselves. Heuristic evaluation is widely used since it is relatively easy and cheap to conduct on software during development as well as on the end product, compared with other evaluation methods such as formal usability testing that involves the use of devices or software to record user behavior in a lab session.

Despite its advantages, there can be problems with heuristic evaluation. During the review, the evaluators have to inspect each screen of the user interfaces to identify problem spots. When the evaluators are not trained for usability evaluation or have less experience in design for usability, they might miss the problems. When the guidelines are general and broad, the evaluation is subject greatly to the evaluators' interpretation of the guidelines. When the guidelines are specific and detailed, there might be a long list of guidelines that the evaluators have to go through. This can be inconvenient and some design problems may be neglected. We aim to assist the evaluators by automating the evaluation task against a number of design guidelines for Android applications which are taken from [2]. In this previous work, we compile a list of design heuristics that comprise general usability guidelines and specific guidelines for mobile applications, as well as a number of newly introduced guidelines specific to user interfaces on the Android platform. In this paper, we present a development of a usability evaluation tool by inspecting source code and reporting locations in the code where violations of the guidelines are found. Since by nature, heuristic evaluation is subjective and still greatly relies on judgment and experience of the evaluators, we cannot completely automate the evaluation but merely help with certain parts of the guidelines where automation of the review is possible. Even so, our experiment suggests that the tool can help save evaluation time and can discover problems that are missed by the evaluators.

Section 2 of the paper gives some background of the work, followed by related researches in Sect. 3. Section 4 describes the method and tool to automate usability evaluation and Sect. 5 presents an experiment. Section 6 concludes the paper.

2 Background

2.1 User Interface Design Guidelines

Several similar lists of guidelines on user interface design exist [1, 3, 4] but the one given by Nielsen [1] is among the very first and widely-referenced. It is a list of ten general rules that can be applied to any user interface design. They are (1) Visibility of system status, (2) Match between system and the real world, (3) User control and freedom, (4) Consistency and standards, (5) Error prevention, (6) Recognition rather than recall, (7) Flexibility and efficiency of use, (8) Aesthetic and minimalist design, (9) Help users recognize, diagnose, and recover from errors, and (10) Help and documentation.

More specific sets of guidelines later emerge for specific use. Particularly for Android applications, Google suggests 17 principles [5]. They are (1) Delight me in surprising ways, (2) Real objects are more fun than buttons and menus, (3) Let me make it mine, (4) Get to know me, (5) Keep it brief, (6) Pictures are faster than words, (7) Decide for me but let me have the final say, (8) Only show what I need when I need it, (9) I should always know where I am, (10) Never lose my stuff, (11) If it looks the same, it should act the same, (12) Only interrupt me if it's important, (13) Give me tricks that work everywhere, (14) It's not my fault, (15) Sprinkle encouragement, (16) Do the heavy lifting for me, and (17) Make important things fast. Even though these guidelines aim for the more recent mobile technology, the essence of the basic general principles by Nielsen still applies.

2.2 Heuristic Evaluation

Heuristic evaluation [1] is a usability inspection method to identify usability problems in user interface design and it can be used in every phase of software development life cycle. An evaluator first uses the application to get the feel of it, and then inspects specific parts of the user interfaces and records the results in an evaluation form, specifying the locations of the problems, the design guidelines that are violated, and severity rating. The evaluation is normally conducted by a team of evaluators and their evaluation results are aggregated. Since the evaluation requires experience, the evaluators with formal usability training would be ideal. Nevertheless, it is often the case that software designers or developers in a software project have to also review the user interface design.

2.3 Android Application Development

Android is an open source Linux-based operating system for mobile devices. It was developed by the Open Handset Alliance, led by Google, and other companies [6].

Android applications are developed in Java by using Android Software Development Kit. Every Android application runs in its own process on its own instance of the Dalvik virtual machine.

The Android architecture consists of four main layers. They are (1) Linux Kernel which handles hardware drivers and networking, (2) Libraries and Android Runtime which provide general libraries (e.g. for web browsers, databases, Internet security etc.) and Android-specific Java-based libraries, together with the Dalvik VM, (3) Application Framework which provides Java classes as higher-level services to applications, and (4) Applications which is the top layer where all Android applications are installed.

Four building block classes are used to build an Android application. They are (1) Activity which is the controller of the user interface and handles user interaction with the smart phone screen, (2) Service which handles background processing associated with the application, (3) Content Provider which handles data and database management, and (4) Broadcast Receiver which handles communication between Android OS and the application. These four building blocks are coupled loosely by the application manifest file "AndroidManifest.xml" that describes each of them and how they interact.

3   Related Work

Due to the broad general nature of existing user interface design guidelines, several researches propose a list of specific guidelines for specific contexts. For example, Gómez et al. [7] compile heuristic evaluation checklists found in the literature and adapt them to mobile user interface. There are 13 design heuristics plus sub-heuristics listed as evaluation questions. That is, there are 158 general and 72 mobile-specific questions. Omar et al. [8] propose a heuristic evaluation checklist for the ERP application on mobile devices. The results are 460 design guidelines.

For Android applications, we use the list of guidelines from Thitichaimongkhol and Senivongse [2]. We adopt their 12 design heuristics which are (1) Visibility of System Status, (2) Match between System and The Real World, (3) User Control and Freedom, (4) Consistency and Standards, (5) Error Prevention, (6) Recognition Rather Than Recall, (7) Flexibility and Efficiency of Use, (8) Aesthetic and Minimalist Design, (9) Help Users Recognize, Diagnose, And Recover from Error, (10) Help and Documentation, (11) Pleasurable and Respectful Interaction, and (12) Privacy. Under these heuristics are 146 evaluation questions.

The advantages of the approach taken by the researches above are that the evaluators can better and directly relate the specific guidelines to the user interface design at hand during the review. Problem areas can be spotted faster, and such detailed lists are useful even for untrained evaluators. On the other hand, the long lists can be intimidating and inconvenient for the evaluation in practice, and there are no automated tools provided by these researches. Nevertheless, a support tool is proposed by Sivaji et al. [9] to help with heuristic evaluation. Specifically, the tool supports the evaluation process by allowing the evaluator to directly log the defects found, and view the defects being captured by another evaluator to minimize and remove duplicates. This tool does not support automated evaluation for heuristic violation though.

4 Automating Heuristic Evaluation: Method and Tool

We attempt to help the evaluators to locate design problems in the user interface of Android applications by automating the evaluation with respect to a number of design guidelines. The application code will be examined to find heuristic violation. The method and accompanying tool are discussed as follows.

4.1 Identify Usability Design Guidelines for Automation

As mentioned previously, we adopt the list of user interface design guidelines for Android applications from [2]. So first, we examine their design guidelines that comprise 146 evaluation questions under 12 design heuristics. Some of the questions apply to user interface design in general, whereas some are specific to design for Android. Additionally, we consider technical information on Android application development in order to determine which evaluation questions have the potential to be checked automatically. It is found that, although this long list of questions gives specific details of how to design the user interface in various situations, the evaluation questions are still subjective and require judgment of the evaluators to answer. Most of the evaluation questions cannot be checked automatically or are difficult to do so due to possibly varying techniques and styles of coding. Table 1 lists an example of the evaluation questions that are not checked automatically by the tool. Table 2 lists 19 questions that we see the potential for

Table 1 Example of evaluation questions not checked automatically by tool

Evaluation questions without automated checking
Evaluation question: Is the logo meaningful, identifiable, and sufficiently visible?   **Design heuristic**: Visibility of system status   **Level of guideline**: General
Evaluation question: If a dialog is showing, can the user be dismissed by touching any area outside the dialog?   **Design heuristic**: User control and freedom   **Level of guideline**: Specific to Android
Evaluation question: Are prompts brief and unambiguous?   **Design heuristic**: Help users recognize, diagnose and recover from errors   **Level of guideline**: General
...

Table 2 Evaluation questions automated by tool

#	Evaluation questions with automated checking
1	**Evaluation question**: Are operating system's status bars mostly (or always) visible, except for multimedia content?
	Design heuristic: Visibility of system status
	Level of guideline: Specific to Android
2	**Evaluation question**: Are operating system's buttons (e.g. back button, home button) mostly (or always) visible, except for multimedia content?
	Design heuristic: Visibility of system status
	Level of guideline: Specific to Android
3	**Evaluation question**: Can operating system's buttons (e.g. back button, home button) be used without blocking by the system?
	Design heuristic: User control and freedom
	Level of guideline: Specific to Android
4	**Evaluation question**: Can every screen in the system be displayed consistently with all devices of the same device type (smartphone, tablet)?
	Design heuristic: Consistency and standard
	Level of guideline: General
5	**Evaluation question**: Has a heavy use of all uppercase letters on a screen been avoided?
	Design heuristic: Consistency and standards
	Level of guideline: General
6	**Evaluation question**: Is there consistent typography across the system?
	Design heuristic: Consistency and standards
	Level of guideline: General
7	**Evaluation question**: Is there consistent design on physical size (font size, element size) across the screen size, and screen density?
	Design heuristic: Consistency and standards
	Level of guideline: General
8	**Evaluation question**: Are menu titles either centered or left-justified?
	Design heuristic: Consistency and standards
	Level of guideline: General
9	**Evaluation question**: Does the system font appearance (size, typeface) can be changed to be consistent with operating system font appearance?
	Design heuristic: Consistency and standard
	Level of guideline: Specific to Android
10	**Evaluation question**: Do objects on the screen have the size that is easy to touch (about 1×1 cm or 48×48 density-independent pixels)?
	Design heuristic: Error prevention
	Level of guideline: Specific to Android
11	**Evaluation question**: Does the search box have the largest possible size that will fit on the screen?
	Design heuristic: Flexibility and efficiency of use
	Level of guideline: General

(continued)

Table 2 (continued)

#	Evaluation questions with automated checking
12	**Evaluation question**: Does the system support both orientations (horizontal and vertical)?
	Design heuristic: Flexibility and efficiency of use
	Level of guideline: General
13	**Evaluation question**: Are several search boxes with different functionalities not used on the same page?
	Design heuristic: Flexibility and efficiency of use
	Level of guideline: General
14	**Evaluation question**: Does the system provide speech-to-text to support searching?
	Design heuristic: Flexibility and efficiency of use
	Level of guideline: General
15	**Evaluation question**: In a data entry form, can the user move focus from one textbox to another textbox by pressing next on virtual keyboard?
	Design heuristic: Flexibility and efficiency of use
	Level of guideline: Specific to Android
16	**Evaluation question**: Does the system not use too many typefaces? (Typefaces can be used to emphasize the content but many typefaces may make users confused.)
	Design heuristic: Aesthetic and minimalist design
	Level of guideline: General
17	**Evaluation question**: Are cyclical animations avoided?
	Design heuristic: Aesthetic and minimalist design
	Level of guideline: General
18	**Evaluation question**: Are unnecessary moving animations of information (e.g. zoom in, zoom out) avoided?
	Design heuristic: Aesthetic and minimalist design
	Level of guideline: General
19	**Evaluation question**: Can the system be protected or confidential areas be accessed with certain passwords?
	Design heuristic: Privacy
	Level of guideline: General

automation and the tool can support. They are under seven design heuristics, i.e. (1) Visibility of System Status, (2) User Control and Freedom, (3) Consistency and Standards, (4) Error Prevention, (5) Flexibility and Efficiency of Use, (6) Aesthetic and Minimalist Design, and (7) Privacy.

4.2 Design Method to Automate Usability Evaluation

In this step, we design a method to automate the checking for user interface problems against each evaluation question by inspecting the application code. The

Table 3 Evaluation method for evaluation question #1

Method for automated checking

Evaluation question: Are operating system's status bars mostly (or always) visible, except for multimedia content?

Design heuristic: Visibility of system status

Level of guideline: Specific to Android

Evaluation method:

1. Import "AndroidManifest.xml".file
2. Read value from root element "manifest"
3. Read value from element "application"
4. Read value from attribute "android:theme" of element "application"
5. Store values in a generated data transfer object (DTO)
6. Get data transfer object (DTO) to check for value "@android:style/Theme.Holo.NoActionBar.Fullscreen"
7. Display evaluation result on screen

Evaluation criteria:

Attribute "android:theme" of element "application" should have the value "@android:style/Theme.Holo.NoActionBar.Fullscreen" as follows:

<application android:theme = "@android:style/Theme.Holo.NoActionBar.Fullscreen">
</application>

---

criteria that are checked in the code are specified. Due to limitation of space, we present seven evaluation methods, one for each design heuristic as an example. They are shown in Tables 3,4,5,6,7,8, and 9.

Table 4 Evaluation method for evaluation question #3

Method for automated checking

Evaluation question: Can operating system's buttons (e.g. back button, home button) be used without blocking by the system?

Design heuristic: User control and freedom

Level of guideline: Specific to Android

Evaluation method:

1. Import "MainActivity.java" file
2. Read value from Java code
3. Find a string "SYSTEM_UI_FLAG_HIDE_NAVIGATION"
4. Store values in a generated data transfer object (DTO)
5. Get data transfer object (DTO) to check for the value "SYSTEM_UI_FLAG_HIDE_NAVIGATION"
6. Display evaluation result on screen

Evaluation criteria:

There should not be a string "SYSTEM_UI_FLAG_HIDE_NAVIGATION" in "MainActivity.java" file

---

Table 5 Evaluation method for evaluation question #4

Method for automated checking

Evaluation question: Can every screen in the system be displayed consistently with all devices of the same device type (smartphone, tablet)?

Design heuristic: Consistency and Standards

Level of guideline: General

Evaluation method:

1. Import "AndroidManifest.xml" file
2. Read value from root element "manifest"
3. Read value from element "supports-screens"
4. Read value from all attributes in element "supports-screens"
5. Store values in a generated data transfer object (DTO).
6. Get data transfer object (DTO) to check for support screen size values
7. Display evaluation result on screen

Evaluation criteria:

Values of all attributes of element "supports-screens" should be "true" as follows:

```
<supports-screens android:resizeable = ["true"]
            android:smallScreens = ["true"]
            android:normalScreens = ["true"]
            android:largeScreens = ["true"]
            android:xlargeScreens = ["true"]/>
```

Table 6 Evaluation method for evaluation question #10

Method for automated checking

Evaluation question: Do objects on the screen have the size that is easy to touch (about 1×1 cm or 48×48 density-independent pixels)?

Design heuristic: Error Prevention

Level of guideline: Specific to Android

Evaluation method:

1. Import "Activity_main.xml" file
2. Read value from root element "LinearLayout" or "RelativeLayout".
3. Read value from element "button"
4. Read value from attribute names "layout_width" and "layout_height" in element "button"
5. Store values in a generated data transfer object (DTO)
6. Get data transfer object (DTO) to check button size, i.e. width and height, to see if button size is easy to touch
7. Display evaluation result on screen

Evaluation criteria:

Values from attributes "layout_width" and "layout_height" of element "button" should be "1×1 cm." or "48×48 dip". Nevertheless, since visual inspection by human evaluators could determine only an estimate of "1×1 cm.", not exact "1×1 cm.", the tool therefore allows the button size to vary, e.g. by ± 0.5 cm on width and height

Table 7 Evaluation method for evaluation question #12

Method for automated checking

Evaluation question: Does the system support both orientations (horizontal and vertical)?
Design heuristic: Flexibility and efficiency of use
Level of guideline: General
Evaluation method:
1. Import "AndroidManifest.xml" file
2. Read value from root element "manifest"
3. Read value from element "activity"
4. Read value from attribute "android:screenOrientation" of element "activity"
5. Store values in a generated data transfer object (DTO)
6. Get data transfer object (DTO) to check for the value "unspecified"
7. Display evaluation result on screen
Evaluation criteria:
Attribute name "android:screenOrientation" of element "activity" should have the value "unspecified" as follows:
<activity
        android:screenOrientation = "unspecified"
        android:name = "com.example.MainActivity" >

Table 8 Evaluation method for evaluation question #18

Method for automated checking

Evaluation question: Are unnecessary moving animations of information (e.g. zoom in, zoom out) avoided?
Design heuristic: Aesthetic and minimalist design
Level of guideline: General
Evaluation method:
1. Import "MainActivity.java" file.
2. Read value from Java code.
3. Find a string "startAnimation".
4. Store values in a generated data transfer object (DTO).
5. Get data transfer object (DTO) and count how many times the string "startAnimation" is found. It should be no more than 2.
6. Display evaluation result on screen.
Evaluation criteria:
Count the number of times the string "startAnimation" is found in the file, and it should be no more than 2

4.3 Develop Tool

A web application is developed in Java as a heuristic evaluation tool and run on the Strut framework. The overview of the tool is shown in Fig. 1. First, an evaluator enters the following application files as input: (1) AndroidManifest.xml is the Android definition file that describes fundamental characteristics of the application such as package name, components of the application etc., (2) Activity_main.xml describes the layout of the page, i.e. the placement of every component on the

Table 9 Evaluation method for evaluation question #19

Method for automated checking

Evaluation question: Can the system be protected or confidential areas be accessed with certain passwords?

Design heuristic: Privacy

Level of guideline: General

Evaluation method:

1. Import "Activity_main.xml" file
2. Read value from root element "LinearLayout" or "RelativeLayout"
3. Read value from element "EditText"
4. Read value from attribute "android:inputType" of element "EditText"
5. Store values in a generated data transfer object (DTO)
6. Get data transfer object (DTO) to check for the value "textPassword"
7. Display evaluation result on screen

Evaluation criteria:

Attribute "android:inputType" of element "EditText" should have the value "textPassword" as follows:

```
<EditText
      android:id = "@ + id/password"
      android:hint = "@string/password_hint"
      android:inputType = "textPassword"/>
```

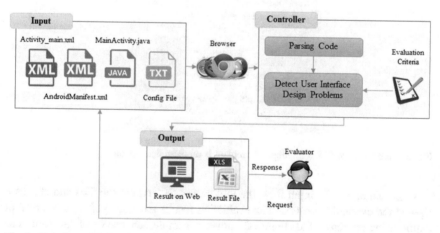

Fig. 1 Overview of heuristic evaluation tool

application screen., (3) MainActivity.java is the actual application file that would be converted to a Dalvik executable and runs the application., and (4) Config file specifies additional information, e.g. which of the 19 evaluation questions are to be checked. In other cases, parameters might be required in a config file such as for evaluation question #11, the name of a search textbox has to be specified, and for question #16, the evaluator can specify how many typefaces are considered too many.

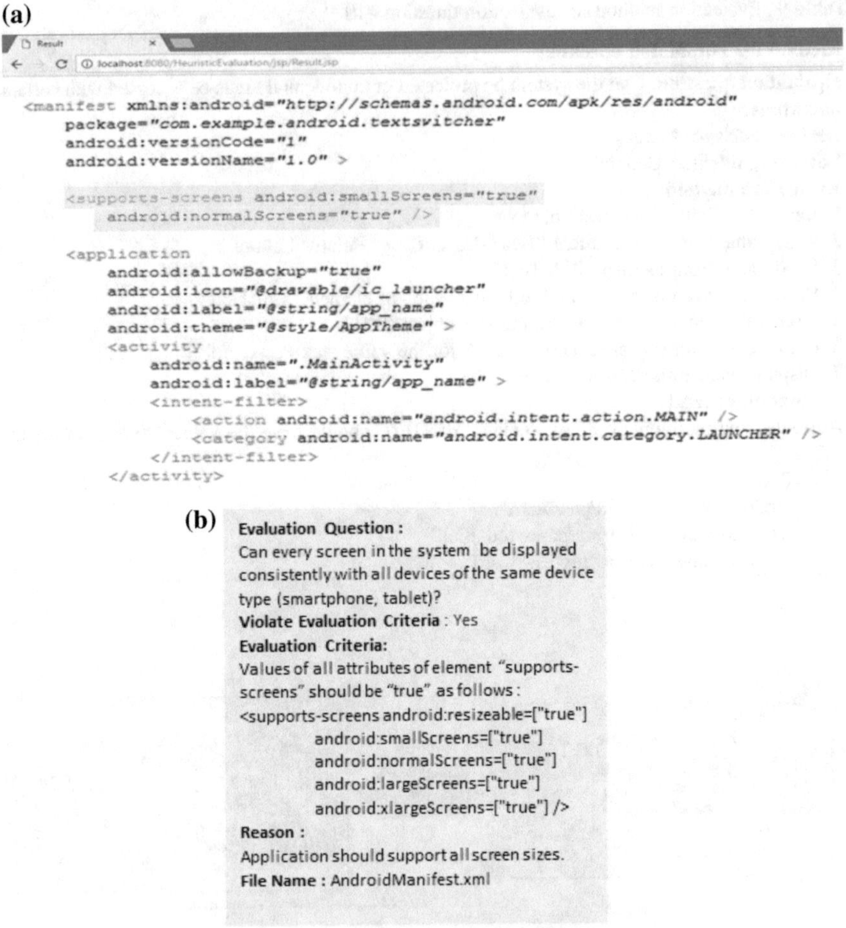
(a)
Result
localhost:8080/HeuristicEvaluation/jsp/Result.jsp
<manifest xmlns:android="http://schemas.android.com/apk/res/android"
package="com.example.android.textswitcher"
android:versionCode="1"
android:versionName="1.0" >
<supports-screens android:smallScreens="true"
android:normalScreens="true" />
<application
android:allowBackup="true"
android:icon="@drawable/ic_launcher"
android:label="@string/app_name"
android:theme="@style/AppTheme" >
<activity
android:name=".MainActivity"
android:label="@string/app_name" >
<intent-filter>
<action android:name="android.intent.action.MAIN" />
<category android:name="android.intent.category.LAUNCHER" />
</intent-filter>
</activity>
(b)
Evaluation Question :
Can every screen in the system be displayed consistently with all devices of the same device type (smartphone, tablet)?
Violate Evaluation Criteria : Yes
Evaluation Criteria:
Values of all attributes of element "supports-screens" should be "true" as follows :
<supports-screens android:resizeable=["true"]
android:smallScreens=["true"]
android:normalScreens=["true"]
android:largeScreens=["true"]
android:xlargeScreens=["true"] />
Reason :
Application should support all screen sizes.
File Name : AndroidManifest.xml

Fig. 2 Screenshot of tool **a** highlighted problem **b** problem description

Once obtaining the input files, the tool parses the code in the files and checks it against the evaluation criteria. The evaluation results are displayed on the screen or exported as an Excel file. Figure 2 shows a sample screenshot of the tool. The problem area within a file is highlighted and the problem description is given.

5 Experiment

In an experiment, we evaluated the tool in terms of evaluation time and problem detection performance. An Android application called "MHMC My Sa-Bai" of a bank in Thailand is used as a test application [10]. It calculates monthly installment

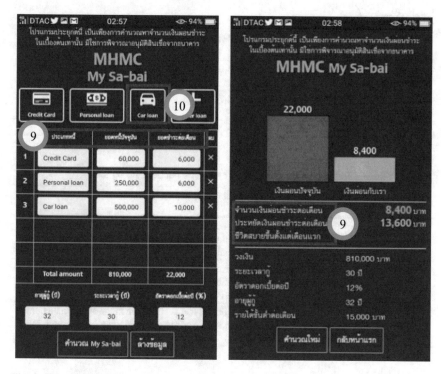

Fig. 3 Test application

for loan and consists of three screens. Sample screens are in Fig. 3. We had six evaluators evaluate the application. They were divided into two groups, three in each group. The expert group have at least three-year experience in user interface design for Android applications. The novice group have one- to two-year experience in developing on Android. The application was checked with the evaluation questions #1–#12 (see Table 2). The evaluators in each group evaluated the application and the results were aggregated. Then the tool was run and problems were reported. Evaluation time was also recorded. Note that, the evaluation time of the tool included time to use the tool (i.e. upload files, execute, display results). The experiment results are in Table 10. For each evaluation question that was violated, one problem was found (at a single location). Even for a small application, the tool could save evaluation time. All of the design problems that were detected by human evaluators were detected by the tool. On the other hand, a number of problems that were detected by the tool were not detected by the evaluators. Discussion about these undetected problems are in Table 11.

Table 10 Results of experiment

Evaluator group	Evaluation questions #Which detect problems	Evaluation questions #Related to undetected problems	Evaluation time (Min.)
Expert group	1,3,5,6,7,12	4,9,10	10
Novice group	1,3,5,6	4,7,9,10,12	20
Tool	1,3,4,5,6,7,9,10,12	–	5

Table 11 Reason for problems undetected by evaluators but detected by tool

Evaluation question#	Reason for undetected problems
4	This test application does not support all screen sizes. It is often the case that evaluators do not conduct evaluation on all screen sizes, e.g. because not all device sizes are available for evaluation
9	Typeface of this test application is not consistent with typeface of Android operating system but the difference does not stand out and could go unnoticed very easily (see Fig. 3)
10	This test application has a button that is smaller than the appropriate button size. This problem goes unnoticed as the button is hidden in between other bigger buttons (see Fig. 3)

6 Conclusion

The proposed tool should be useful for heuristic evaluation by evaluators with formal usability training as well as the less experienced. The tool can help save evaluation time and can detect design problems that can be easily missed by evaluators. Since only 19 evaluation questions are automated, the tool cannot replace human evaluators. The tool is also limited in that it cannot analyze the code that is not available statically in the program such as the code of the user interface libraries. Nevertheless, the evaluators can use the tool to first automatically check if there are any violations to those 19 questions. Then they can proceed with heuristic evaluation as usual especially for the remaining evaluation questions. Additional experiment is planned to evaluate the accuracy of the tool on applications of different sizes and domains.

References

1. Nielsen, J.: Usability engineering, pp. 115–148. Academic Press, San Diego (1994). ISBN 0-12-518406-9
2. Thitichaimongkhol, K., Senivongse, T.: Enhancing usability heuristics for Android applications on mobile devices. In: Lecture Notes in Engineering and Computer Science:

Proceedings of the World Congress on Engineering and Computer Science 2016 (WCECS 2016), pp. 224–229 (2016)
3. Shneiderman, B., Plaisant, C., Cohen, M., Jacob, S.: Designing the user interface: strategies for effective human-computer interaction, 5th edn. Prentice Hall (2009)
4. Pierotti, D.: Heuristic evaluation—system checklist, Technical Report. Xerox Corporation, Society for Technical Communication (1998)
5. Google Inc.: Android design principles. https://developer.android.com/design/get-started/principles.html (2017). Accessed 9 Apr 2017
6. Tutorials Point India Private Limited.: Android development. https://www.tutorialspoint.com/android/index.htm (2017). Accessed 9 April 2017
7. Gómez, R.Y., Caballero, D.C., Sevillano, J.L.: Heuristic evaluation on mobile interfaces: a new checklist. Sci. World J. **2014**, Article ID 434326, pp. 1–19 (2014)
8. Omar, K., Rapp, B., Gómez, J.M.: Heuristic evaluation checklist for mobile ERP user interfaces. In: Proceedings of 7th International Conference on Information and Communication Systems (ICICS), pp. 180–185 (2016)
9. Sivaji, A., Soo, S., Abdullah, M.R.: Enhancing the effectiveness of usability evaluation by automated heuristic evaluation system. In: Proceedings of 3rd International Conference on Computational Intelligence, Communication Systems and Networks, pp. 48–53 (2011)
10. The Siam Commercial Bank Public Company Limited.: "MHMC My Sa-Bai" Application (2017). Accessed 16 Apr 2017

Automated Essay Scoring System Based on Rubric

Megumi Yamamoto, Nobuo Umemura and Hiroyuki Kawano

Abstract In this paper, we propose an architecture of automated essay scoring system based on rubric, which combines automated scoring with human scoring. Rubrics are valid criteria for grading students' essays. Our proposed rubric has five evaluation viewpoints "Contents, Structure, Evidence, Style, and Skill" and 25 evaluation items which are subdivided viewpoints. The system is cloud-based application and consists of several tools such as Moodle, R, MeCab, and RedPen. At first, the system automatically scores 11 items included in the Style and Skill such as sentence style, syntax, usage, readability, lexical richness, and so on. Then it predicts scores of Style and Skill from these items' scores by multiple regression model. It also predicts Contents' score by the cosine similarity between topics and descriptions. Moreover, our system classifies into five grades "A+, A, B, C, D" as useful information for teachers, by using machine learning techniques such as support vector machine. We try to improve automated scoring algorithms and a variety of input essays in order to improve accuracy of classification over 90%.

Keywords Automated scoring · Essay evaluation · Rubric · Cosine similarity · Support vector machine · Multiple regression model

M. Yamamoto (✉)
School of Contemporary International Studies, Nagoya University of Foreign Studies,
Nisshin 470-0197, Japan
e-mail: yamamoto@nufs.ac.jp

N. Umemura
School of Media and Design, Nagoya University of Arts and Sciences,
Nagoya 470-0196, Japan
e-mail: d_chaser@nuas.ac.jp

H. Kawano
Faculty of Science and Engineering, Nanzan University, Nagoya 466-8673, Japan
e-mail: kawano@nanzan-u.ac.jp

© Springer International Publishing AG 2018
R. Lee (ed.), *Applied Computing & Information Technology*,
Studies in Computational Intelligence 727, DOI 10.1007/978-3-319-64051-8_11

1 Introduction

The authors are developing an automated essay scoring (AES) system on the Learning Management System (LMS) to support students' writing skills up and help teachers scoring essays. In the essay evaluation, there are various problems such as variation in grading within one teacher, bias in grading among teachers, the burden of time, and so on. Most of teachers are making efforts to maintain more strict evaluation by specifying checklists or evaluation indicators so-called rubrics. Therefore, it is natural to design algorithms for scoring based on these rubrics for developing the AES system.

In recent years, it is discussed the necessity of active learning and the strict evaluation. It is difficult to evaluate the results of studies in such classes in a conventional examination format. Therefore, performance evaluation like essays, papers, presentations are adopted. Teachers are required to support students for writing and score their essays strictly. In order to solve these problems, we propose a rubric for human scoring and introduce it to the automated scoring system.

In this paper, we report on the architecture of our AES system for teachers, which is the core of the installed system at the present stage.

2 Background

Research on the AES system was started by Page (USA) in the 1960's [1]. Due to the development of natural language processing (NLP), information retrieval technology and corpus, AES systems based on various methods have been developed. These systems are still being improved. In the United States, there are some commercial AES systems already in operation such as e-rater, PEG, IEA, IntelliMetric, BETSY [2]. In recent years, Massive Open Online Courses (MOOCs) also use the AES system in some grades. In Japan, Jess has been developed by Ishioka and Kameda [3]. e-rater, PEG, IEA, and Jess calculate the total score with continuous quantity. On the other hands, IntelliMetric, BETSY classify into score categories. The summary of the features of these AES systems is presented in Table 1. This table follows "Table 1: Comparison of Essay Evaluation System" by Ishioka [4].

For example, e-rater calculates 12 feature values such as grammar, usage, mechanics, style, organization, vocabulary, and so on. Then, it achieves the desired scale as a weighted average of the standardized these feature values, followed by applying a linear transformation. IntelliMetric train with a set of pre-scored essays with known scores assigned by human raters. It has five categories of features, such as focus and unity, organization, development, and elaboration. Jess scores rhetoric, logical composition, and contents at 5: 2: 3. It finds the ideal distribution of features from the description of professional writers, not professional raters. The essays with antlers are scored lower.

Table 1 The comparison of AES system

System	Example of usage	Main focus for evaluation	Rubric[e]	Technique	Notices	Accuracy[f]
e-rater V.2	TOEFL, GMAT[a]	Structure, Organization, Contents	Not used	Multiple regression model using	12 features (grammar, style, usage, mechanics, etc.) used	0.97 [5] (2006)
Intelli-Metric	Bar exam in Pennsylvania MCAT[b]	Focus and Unity, Organization, Development and Elaboration, Sentence Structure, Mechanics and Conventions	Not used	Linear analysis, Bayesian, Latent Semantic Analysis (LSA)	Large amounts of data are required for each topic	0.80–0.85 [6] (2006)
Jess	(Aims to use in) NCUEE[c]	Rhetoric, Organization, Contents	Not used	Statistics (abnormal value detection)	Weak in science and technology fields	0.57 [3] (2006)
MOOCs (edX)	MOOCs[d]	Readability, The number of Characters and words	Not used	Use scoring patterns of 100 essays scored by humans	Not only teachers but also students score grades of others	Not published

[a]The Graduate Management Admission Test
[b]Medical College Administration Test
[c]The National Center for University Entrance Examinations
[d]Massive Open Online Courses
[e]Rubric for Automated Scoring
[f]Correlation with human scoring

3 The Proposal of Rubric as Evaluation Criteria

3.1 Necessity of Rubric

Ishioka discusses the requirements for AES systems and concludes both e-rater V.2.0 and Jess are valid. Since these scores based on the evaluation criteria table, so that they are constant and topics are not limited [2]. The evaluation criteria table he mentioned is equivalent to a rubric. However, it follows the grading standard of GMAT which is an entrance examination to the management school of the United States.

In this research, our aim is essay scoring system which can be generally and practically used at the educational site in university. Therefore, we consider that it should be carefully analyzed and prepared the existing report writing rubric, and then designed the AES system based on it.

3.2 Proposed Rubric

We propose the rubric for evaluation of essays (see Appendices 1 and 2). This rubric is developed with reference to the Written Communication VALUE Rubric [7] in AAC & U of the United States and Common rubric (writing) of Kansai International University, etc. These rubrics are created and utilized at the organizational level in universities. As for the former, we referred to the original text published on the AAC & U website and Matsushita's translation [8]. Table 2 shows the comparison with the reference rubric.

Proposed rubric has five types of evaluation viewpoints (Contents: validity of task understanding and content of the answer, Structure: Logical development, Evidence: validity of documentation and evidence, Style: compliance and appropriate elaboration of sentence writing method, Skill: skill of reading and expression). Since these viewpoints are too ambiguous to be calculated, they are divided into 25 items. Thus, we developed a kind of the automated scoring rubric. These items and expressions are consulted them that are common in the report scoring rubrics published on the web such as rubric's paper or so-called rubric bank. We consider the items 1, 2, 14–25 in the appendix rubric can be applied to automatic scoring. At present, the items 2 and 14, the automatic scoring is scheduled, but the implementation has not completed yet.

3.3 Scoring Model Based on Rubric

Content, Structure, and Evidence (abbreviated as "CSE") of the evaluation viewpoints strongly depend on the judgment of the teacher's value, so it is difficult to automatically score all items. CSE positions the contents "1: Similarity between topic and description" and "2: Presence of keywords" as automated scoring items at present.

On the other hand, since the evaluation items of Style, Skill (abbreviated as "SS") are measurable quantitative data, they can be subject to automated scoring. As for the evaluation of SS, automated scoring is accuracy and preferable rather than human scoring because it can be avoided differences in evaluation among teachers. After all scores of SS are calculated, each final score of SS is calculated by multiple regression models with these scores as feature vectors.

Moreover, based on the results of the above 13 automated scoring items, the final score and the level of the essay are estimated by the classifier. Specifically, a classifier for predicting and classifying the final score and level is prepared by

Table 2 The comparison of rubrics

Rubric (Developer)	Evaluation viewpoints		Evaluation levels		
	#1	Contents	#2	Contents	Notes
Written Communication VALUE Rubric (AAC&U)	5	Context of and purpose for writing, Content development, Genre and disciplinary conventions, Sources and evidence, Control of syntax and mechanics	5	0 (does not meet benchmark), 1 (Benchmark), 2 and 3 (Milestones), 4 (Capstone)	Abstractly created for localization. (Change according to each university and lesson)
Common rubric (writing) (Kansai University of International Studies)	5	Relevance to the topic, Logical configuration, Reference, Style elaboration, Writing skill	5	Level 0–4	Created separately for low grade and high grade. Change achievement level. Add viewpoints according to subjects
Proposed rubric	5	Content, Structure, Evidence, Style, Skill	10	A+, A, B, C, D (2 points for each level)	Subdivide evaluation items for each viewpoint

#1: the number of viewpoints
#2: the number of levels

Support Vector Machine (SVM) which is trained using feature vectors. The results of this score and level are presented to teachers. This scoring model is shown in Fig. 1.

The teacher can focus on scoring the items of CSE, such as the logical development, structure, and evidence. The final judgment may be made while referring to the overall evaluation value presented by the system. Since the SS items of the essay include information on the readability of the sentences and writing skills, it also greatly influences the teacher's reading of the contents. Therefore, we consider that it is appropriate to obtain the overall score by using the results of SS as explanatory variables.

Evaluation Items based on rubric

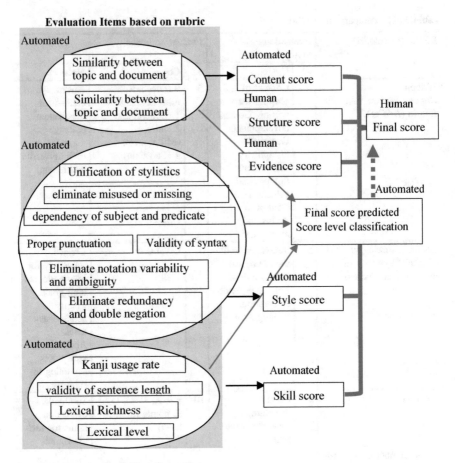

Fig. 1 The scoring model based on rubric

4 System Configuration

4.1 Outline of the Proposed System

Figure 2 shows the architecture of our proposed system. This system is the web based application, students submit essays in text format to LMS. Teachers can view the scoring results of all essays in class, from "grading process for faculty" provided on the LMS.

We have already implemented simple text mining plugin (TeMP) on LMS. TeMP [9] calculates the basic statistical information of the text, such as the number of characters, the number of sentences, and the type-token ratio which are needed to score SS. Therefore, our system is developed as the extension function of TeMP.

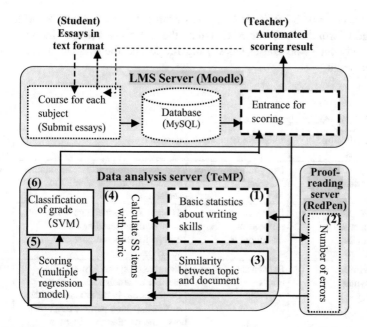

Fig. 2 The architecture of our proposed system

The modules (1), (3)–(6) of the scoring process are programs in PHP or R. Module (2) uses RedPen server which is an open source proofreading tool.

The main feature is that the scoring items are determined based on the rubric which is for human scoring. We also use machine learning techniques such as cosine similarity, multiple regression model, and SVM.

4.2 Process of Automated Scoring

After students submit their essays on the LMS, the teacher selects that data to be processed and starts to analyze. The system presents the scoring result through the following six processes.

(1) After morphological analysis of documents, extract nouns, verbs, and adjectives and create a term-document matrix showing the appearance frequency of index terms. Calculate basic statistics such as the number of characters per document, the number of sentences, and so on.

(2) Obtain the number of errors from the proofreading server RedPen [10]. Sum them for each document and for each evaluation item.

(3) Calculate the cosine similarity between the topic and essays, based on the vector space model.

(4) Each value of SS items is calculated from (1) and (2) according to the calculation formulae in Table 3. Calculate the difference between the proper value already set as an automated scoring rubric and the evaluation value of each document.

Table 3 The scoring algorithm of each item

No	Evaluation items	Judgment contents	Formula	Proper value
15	Unification of stylistics	Unity in 'Normal' or 'Respect'	The non-uniform rate = the number of non-uniform/the number of sentences	0.0
16	Eliminate misused or missing characters	Existence of misused or misspellings	The rate of misspellings = the number of misused and misspellings/the number of sentences	0.0
17	Validity of syntax	Existence of syntax errors	The rate of syntax errors = the number of ambiguous noun Conjunction, in case clause too deep or use the same particle more than once/the number of sentences	0.0
18	Dependency of subject and predicate	Validity of Relationship between subject and predicate	Twist or confusion rate between subject and predicate = number of duplicate of incidental particle/the number of sentences	0.0
19	Proper punctuation	Average number of characters between punctuations	The average number of characters between punctuations = total number of characters/the number of punctuations	11–15
20	Eliminate redundancy and double negation	Double words	The rate of redundancy and double negation = the number of redundancies and double negation/the number of sentences	0.0
21	Eliminate notation variability and ambiguity	Notation, variability, and ambiguity	The rate of notation variability and ambiguity = the number of notation variability and ambiguity/the number of sentences	0.0
22	Kanji usage rate	Kanji usage rate	Kanji usage rate = number of Kanji/total number of characters	>32%
23	Validity of sentence length	Average of sentence length	Average of sentence length = total number of characters/the number of sentences	26–41
24	Lexical Richness	Type-Token Ratio	TTR = Types/Tokens	1.0
25	Lexical level	Average of lexical level point	Average of lexical level point = each lexical level of all nouns, verbs, and adjectives*each weighting	6.0

(5) Normalize the evaluation value of (4) and calculate the final score of SS by the weighting previously obtained from the learning data.

(6) Using the results of automated scoring, the comprehensive score is estimated by SVM and classified as the score level of "A+" to "D".

4.3 Algorithms of Automated Scoring

As mentioned in Sect. 3.2, we consider the items 1, 2, 14–25 in the appendix rubric can be applied to automatic scoring, and implement the automatic calculation process using eleven items of SS items 15–25 among them. The scoring algorithms of them are as shown in Table 3.

"15 Unification of stylistics": The RedPen server returns the number of places where the style is not unified as the number of errors. The error rate is obtained by dividing that value by the number of sentences. Since the error rate 0.0 means that there is no error, the evaluation point will be the highest score 1.0. The error rate 1.0 means that every sentence has errors, so the evaluation point will be the lowest score 0.0. The other SS evaluation items, 16–18, 20, and 21 are obtained in the same way. "19 Proper punctuation" is judged by the average number of characters between punctuations. This is compared with the appropriate value set for rubric for automated scoring.

For "22 Kanji usage rate" and "23 Validity of the sentence length", the evaluation value is obtained based on the difference from the set appropriate value. These appropriate ranges are based on the values indicated in the grammar proofing tool. "24 Lexical Richness" is judged by type-token ratio.

"25 Lexical Level" is calculated using the Japanese language education vocabulary level table named "jReadability" that Sunakawa (2012) provides as research results [11]. They developed a list which each vocabulary is appended with "1. Elementary Level" to "6. Advanced Level". All index terms in the matrix of process (1) are assigned the level from "jReadability", multiplied the level by the occurrence frequency and calculated average.

"1. Similarity between topic and document" is scored in the process (2) and (3) in Fig. 2. In the process (2), it is created the term frequency vectors of each document (topic and description). The cosine similarity of these vectors is calculated as the similarity of topic and description of the essay.

4.4 Automated Scoring of the Final SS Scores

In order to calculate the final SS scores, we use a regression modeling approach. 80 samples are used as training data. Each evaluation item in Table 3 is automatically scored and used as an explanatory variable. On the other hand, a teacher scores the evaluation viewpoints SS using rubric for human scoring in appendix rubric. We calculate the weights as the explanatory variable for the automated scoring of items and the target variable for the human scoring. Each weight for Style (#15–#21) is 0.299, 1.192, 2.281, −0.701, −0.083, −0.390, −0.316, and 7.194. As for Skill (#22–#25), each weight is 2.624, −0.009, −0.527, 0.894, and 2.578. Therefore, we can see #18, #21, #22 have many influences on this weighting, the final SS scores are calculated.

4.5 Classification

In order to set up classifier, we use short essays submitted by 83 freshmen who take "information literacy subjects" class. The topic of the essay is "data analysis using spreadsheet Excel", the average number of characters is 378.4, the standard deviation is 202.8. The process of setting up the classifier as the preparation for classification is as follows:

1. select 55 essays which are two-third of the 83 essays
2. use them as training data, and others as test data
3. try various classifiers such as SVM and Decision tree
4. choose one of them and save it as the classifier for classification.

As a result of, we adopted SVM's Gaussian kernel with the lowest misclassification rate.

5 Evaluation of the AES System

5.1 Accuracy of Scoring Result

Table 4 shows the characteristics of scoring essays and the correlation between human scoring and automated scoring. The essays are submitted by freshmen in university. The students attend information literacy class, the topic is a consideration of data analysis using spreadsheet software.

In this experiment, we confirmed the correlation of the evaluation viewpoint Style. On the other hand, the correlation with the evaluation viewpoint Skill is low.

Table 4 The spearman correlation with human score

Class	Number of essays	Average number of characters	Standard deviation of characters	Correlation with human score		
				Style	Skill	Classification of score level
A	43	427.6	216.0	0.602	0.210	0.581
B	40	325.5	171.7	0.463	0.089	0.550

Table 5 The accuracy of classifier

Classifier		Incorrectly classified rate	
		Training data	Test data
SVM	Gaussian (rbfdot)	10.9	46.4
	Linear (vanilladot)	34.5	42.9
	Polynomial (polydot)	27.3	53.6
Decision tree		23.6	42.9

Therefore, we are going to improve the calculation algorithm for evaluation items of viewpoint skill.

5.2 Accuracy of Classification

The predicted final score and overall grade levels (A, A+, etc.) are presented when the teacher determines the final score. At the present, we create classifiers exploratory. We utilize multiple kernels on SVM. We also use decision trees, as we visualize the most influential items.

Table 5 shows these classification error rates. It can be seen that the error rates of the training data and the test data are different. From these results, we consider that it is necessary to increase the number of training data and types of levels. At present, we try to improve the accuracy of classification over 90% (error rate within 10%).

6 Conclusion

In this paper, we propose the architecture of AES system based on rubric, and explain the outline of the system under construction. In our proposed system, evaluation items based on rubric are classified into human scoring and automated scoring. In addition, we predict the final score and the grade level from the automated scores using the machine learning to support teachers' scoring.

Currently we are advancing the following points:

- Create the user interface to display scoring results
- Evaluation of processing speed by test data
- Improvement of scoring algorithm by automatic evaluation items
- Improvement of classification accuracy of scoring results by machine learning algorithm.

Appendix 1: Proposed Rubric for Human Scoring

Evaluation Viewpoint	Achievement Level and Scoring				
	D (0–1)	C (2–3)	B (4–5)	A (6–7)	A+ (8–9)
[Content] Understanding of the assigned tasks and validity of contents	Misunderstanding the assigned task, or the contents are not related to the topic at all	Understanding the assigned task, but includes some errors	Understanding the assigned task, but the contents are insufficient	Understanding the assigned task, but has some points to improve	Appropriate contents with relevant terms. No need for improvement
[Structure] Logical development	No structure or theoretical development	There is a contradiction in the development of the theory	Although developing theory in order, there are some points to be improved	Although developing theory in order, the theory is not compelling	The theory is compelling and conveying the writer's understanding
[Evidence] Validity of sources and evidence	It does not show evidence	Demonstrates an attempt to support ideas	The sources to be referenced are inappropriate or unreliable	Uses relevant and reliable sources, but the way of reference is not suitable	Demonstrates the skillful use of high-quality and relevant sources
[Style] Proper usage of grammar and elaboration of sentences	There are some grammatical errors. Many corrections required	Not following the rules. Some corrections required	Almost follow the rules. A few corrections required	Although error-free, some improvement will be better	Virtually error-free and well elaborated. No point to improve
[Skill] Readability and writing skill	The sentences are hard to read. Writing skills are missing	There are several points to be improved, such as the length of sentences	Although sentences can be read generally, some improvement will be better	Easy to read. Rich in vocabulary	Easy to read. Skillfully communicates meaning to readers. Rich in vocabulary

Appendix 2: Proposed Rubric for Automated Scoring

Evaluation Viewpoints	Evaluation Items		Automated Scoring (0–9)
[Content]	1	Similarity between topic and description	Applicable
	2	Presence of keywords	Applicable
	3	Understanding of the writing task	Not applicable
	4	Comprehensive evaluation of contents	
	5	Understanding of learning contents	
[Structure]	6	Logic level	Not applicable
	7	Validity of opinions and arguments	
	8	Division of facts and opinions	
	9	Persuasiveness	

(continued)

(continued)

Evaluation Viewpoints	Evaluation Items		Automated Scoring (0–9)
[Evidence]	10	Quality level of reference material	Not applicable
	11	Relevance of reference material	
	12	Validity of reference material	
	13	Explanation about tables and figures	
	14	Validity of the quantity of citations	Conditionally applicable
[Style]	15	Unification of stylistics	Applicable
	16	Eliminate misused or misspellings	
	17	Validity of syntax	
	18	Dependency of subject and predicate	
	19	Proper punctuation	
	20	Eliminate redundancy and double negation	
	21	Eliminate notation variability and ambiguity	
[Skill]	22	Kanji usage rate	Applicable
	23	Validity of sentence length	
	24	Lexical richness	
	25	Lexical level	

References

1. Shermis, M. D., Burstein, J.: Handbook of Automated Essay Evaluation: Current Applications and New Directions. Routledge, pp. 1–353 (2013)
2. Ishioka, T.: Latest trends in automated essay scoring and evaluation. Trans. Jpn. Soc. Artif. Intell. 23(1), 17–24 (2008) (in Japanese)
3. Ishioka, T., Kameda, M.: Automated Japanese essay scoring system based on articles written by experts. In: Proceedings of the 21st International Conference on Computational Linguistics and 44th Annual Meeting of the ACL, pp. 233–240 (2006)
4. Ishioka, T.: Computer-based writing tests. J. Inst. Electron. Inf. Commun. Eng. 99(10), 1005–1011 (2016) (in Japanese)
5. Attali, Y., Burstein, J.: Automated essay scoring with e-rater® V.2. J Technol. Learn. Assess. 4(3), 3–30 (2006)
6. Vantage Learning: Research Summary IntelliMetric™ Scoring Accuracy Across Genres and Grade Levels. www.vantagelearning.com/docs/intellimetric/IM_ReseachSum-mary_IntelliMetric_Accuracy_Across_Genre_and_Grade_Levels.pdf
7. Association of American Colleges and Universities: Inquiry and analysis VALUE rubric. www.aacu.org/value/rubrics/inquiry-analysis

8. Matsushita, K.: Assessment of the quality of learning through performance assessment: based on the analysis of types of learning assessment. Kyoto Univ. Res. High. Edu. **18**, 75–114 (2012). (in Japanese)
9. Yamamoto, M., Umemura, N.: Analysis and Evaluation of Reports based on Lexical Richness. In: Moodle Moot Japan 2015 Proceedings, pp. 6–8 (2016) (in Japanese)
10. Recruit Technologies Co., Ltd.: RedPen. redpen.cc/
11. Sunakawa, Y., Lee, J., Takahara, M.: The construction of a database to support the compilation of Japanese learners dictionaries. Acta Linguistica Asiatica **2**(2), 97–115 (2012)

Mobile Development Tools and Method Integration

Mechelle Grace Zaragoza, Roger Y. Lee and Haeng-Kon Kim

Abstract As to providing a common framework for a sound generic framework methods and integration tools, software development and distribution support management in the development process is needed in order to meet the required standards of mobile development. This paper supports technical software development activities, integrated software management, and software distribution management. It is used in the MTICASE environment (Method Tool Integration CASE) for the development of software distribution, and is extended to the configurations of heterogeneous components programmed in the programming languages differently. This article shows how a similar set of principles, practices, and programming tools can be combined with recent work by MetaObjects to provide a framework for methods and tools as well as configuration.

Keywords Configuration MetaObject · Framework · Application domain

1 Introduction

Mobile development is everything as Mobile phones have become part of many people's daily life. According to the International Telecommunication Union, 6.8 billion mobile phone users worldwide at are expected to appear by 2012, which is equivalent to 96% of the entire world population and a big leap from 5.4 billion mobile phone users in 2010 [1]. To as software component, it is simply impossible to distinguish from other software elements of the programming language. The difference should be how the software components are used. The software includes

M.G. Zaragoza · H.-K. Kim (✉)
Catholic University of Daegu, Gyeongsangbuk-do, South Korea
e-mail: hangkon@cu.ac.kr

M.G. Zaragoza
e-mail: mechellezaragoza@gmail.com

R.Y. Lee
Computer Science Department, Central Michigan University, Mount Pleasant, USA

© Springer International Publishing AG 2018   191
R. Lee (ed.), *Applied Computing & Information Technology*,
Studies in Computational Intelligence 727, DOI 10.1007/978-3-319-64051-8_12

a set of abstract qualitative characteristics, namely, abstract, quality features, that is, the extent to which an element or process meets the requirements of (IEEE Std 610.12–12 1990).

A **software component** is a software element that corresponds to a component model and can be deployed independently, and the connection has no change in the composition standard.

The **component model** defines the interaction and species its composition standards.

An **infrastructure software component** is a set of interactive software components designed to ensure that the subsystem system or software.

Software metrics related to program attributes such as lines of code, complexity, frequency of modification, coherency, coupling, etc [2].

Following definitions demonstrate the important relationship between software for infrastructure components, software components and component of the model [3].

Moreover, even in a one-step process is necessary to represent the different aspects of the application, not only in terms of the area of application of the score, but for the provision of various shows such as functionality, performance, fault tolerance, security and others. There is a need for other methods and symbols, as well as the possibility of modification and integration. The concept is the basis for the integration of methods that supports distributed development teams and staff provide instrumental support [4]. Software reliability engineering focuses on engineering mechanisms for quantitative evaluations of software reliability, the development of software, and the maintenance of software [5]. Program analysis is the process of analyzing the behavior of a computer program. Program analysis has a very widely application range, it provides support for compiler optimization, testing, debugging, verification and many other activities [6]. In addition, even at a single stage in the process, it is necessary to represent different aspects of the application, not only in terms of a partitioning of the application domain but also to provide different views such as functional, performance, fault tolerance, safety and others.

1.1 Addressing Meta Objects

The study addresses these issues and provides a customizable framework for integrating methods and tools. This approach is based on the application of domain decomposition, notation and process steps. The main objects that are configured in this structure are MetaObjects. A method is described as a configuration of "interacting" MetaObject types (templates). Method use involves MetaObject instantiation to provide the specifications in each notation. Method integration forms the software configuration process at MetaObject level. Each MetaObject can be elaborated separately to interacte the several constraints. Tool support can be provided for each type of MetaObject and for configuration and interaction of these constraints.

1.2 Configuring Software

Since the web environment is easy to access and provides immediate feedback on software modifications, software in this environment is very attractive to the companies that attempt to reduce development costs and respond to rapid technological advance [7]. The rapid development and quality of orderly software is demanding and difficult. Software systems large size and complex development and maintenance are challenging to manage. Complex software systems such as software systems for the space shuttle, or air traffic control are among the most complex ever built. They are the fruit of a collaboration of many people who worked together as a team, with each person or group controlling only a small part of the whole system. Therefore, we need methods and mechanisms to divide a large system into manageable parts, and even more important to put the parts [8]. The analogy to configure of MetaObjects and their interactions is noticeable. We see MetaObjects as the means for encapsulating each aspect and notation, and configurations as the means for gluing them together. The description of method and process structure as a configuration is separated from the description of a MetaObject.

Complex MetaObjects can also be defined as compositions of simpler ones. We believe that the configuration level provides a convenient level of abstractions at which we can view and manage methods and software projects in a different way. In addition, its declarative structural form makes it independent of the procedural aspects.

2 Basic Principles

The basic principles of the configuration programming approach are summarized:

2.1 Separate Structural Description of the Basic Component

Separation of concerns facilitates the description, comprehension and manipulation, by both man and machine. This is achieved by abstraction away from the component programming concerns. The structure of the configuration specification makes it agreeable to both textual and graphical description. System construction can be performed by translation of the structural configuration description by component creation and interconnection. In addition, the configuration language should be declarative, describing what the structure is, not how it is to be constructed [9].

Declarative descriptions tend to be more concise and amenable to analysis, interpretation, and manipulation than their imperative equivalents.

2.2 Components

As to context independence, it defined as the component that makes no direct reference to any non-local entities, but can be integrated into any compatible context w/out redefining or recompiling it. We therefore require that components access only local data and use indirect naming to refer to connected components.

Definition as a type permits instantiation and reuse in different contexts. The component interface should describe the interaction points with other components and permits validation of interconnections at configuration time (see Fig. 1).

2.3 Complex Components

Hierarchies are a natural and convenient means to support of sub component encapsulation and information in Fig. 2.

Interconnected instances of more basic component types can be composed to form more complex components e.g. (instance hierarchy). These composite components should be component types that are available to use in further definitions; such an approach also permits the definition and construction of recursive structures.

Fig. 1 A framework for presenting the state-of-the-art in software engineering research for mobile apps [13]

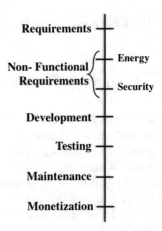

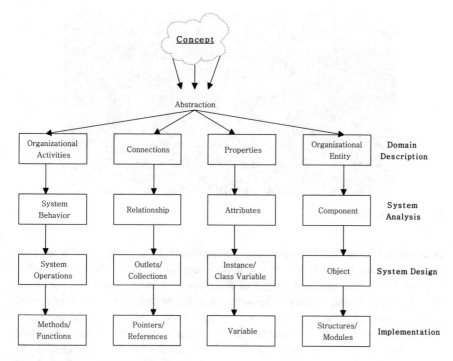

Fig. 2 Component abstraction evolution

3   MetaObject

3.1   MetaObject Definition

MetaObjcets is the calculation of the target object as to manipulates, create or implement a description of the objects [10]. Thus, MetaObject is a combination of the following parts which we refer to as slots:

1. A style, the representation scheme in which the target object expresses what you can see (examples of types of data flow analysis, modeling entity relationship attributes, Petri nets, logical equations, etc.);
2. A domain defines which part of the "world" is defined in the way (how style is a structured representation defined) by the target object (for example, control system would lift areas like the user, the elevator and controller);
3. A specification, expressed in the kind metaobject's outline claims specific areas;
4. A work plan, how and under what circumstances the content of the specification is deliberate and modified;
5. A work folder, a summary of the current state of development.

The Metaobject in Fig. 3, presented in encapsulated knowledge in the form of different slots, style and specification. Style slots and work plan are generally

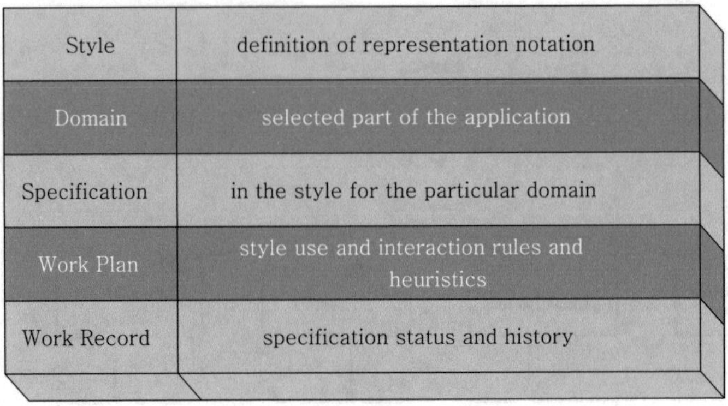

Style	definition of representation notation
Domain	selected part of the application
Specification	in the style for the particular domain
Work Plan	style use and interaction rules and heuristics
Work Record	specification status and history

Fig. 3 A Meta Objects

encapsulated in the field of slots. The description and the inclusion of a working subject provides a specific knowledge to do a particular problem. The specification is given in a single coherent style and describes an identified problem area field. This also includes the interaction between metaobjects transferring information and activities such as consistency checks. If a target specific object is to be used on a data flow diagram in the specification site for a particular range, other slots would be required to use the language specified in the DFD (data flow diagram) representation [11].

3.2   Discussed Principles to MetaObjects

We separate the languages used to describe the structure in the MetaObjects. This physical structure is useful for understanding the description and manipulation of MetaObjects. The configuration language is declarative, describing what the structure is, not how it is to be constructed. We refer to these as templates (Fig. 2). A MetaObject template consists of a MetaObject in which only the style and the work plan have been defined. Context independence means that the workplan makes no direct reference to any non-local entities. Mappings to other MetaObject interface describes the interaction points with other MetaObject. By using the configuration language, complex MetaObjects should be definable as a composition of instances of MetaObject templates.

The style slot of a complex MetaObject permits the use of the configuration language to specify it as a configuration. Change should be expressed at the configuration level, as changes of the MetaObject instances and/or their interconnections.

Modifications to the method and specification are reflected as changes to the configuration.

3.3   Configuration Framework

3.3.1   MetaObject Templates

A metaobject template only elaborates the work plan style slot. These aspects are closely related as the work plan. It describes basic actions that need to be performed to provide a specific given style. These actions are general. It can be used to specify the source and selected stuff from the application domain. Such a specification is termed a MetaObject instance since it refers to a specific instantiation of the template, and would include identification of the selected domain and elaboration of the specification. A method is described as a form of configuration of a selected set of MetaObject templates which together describe the styles and work plans to be used in the method. The mappings and checks between templates should also be specified. The dynamics of the method are described by permitting one MetaObject to create another as the method unfolds. Method use is thus represented as a dynamically evolving configuration of MetaObjects.

An example of a model for the initial stage fare model is given in Fig. 3. Models of the World Real effect, a style called SSD (System Schema Specification). This DESCRIPTION provides an overview of the constraints for a "well-formed" style diagram. The description of the actions in the work plan describes building as a specification the solid state, along with any restrictions desired in ordering actions fare work plan.

- Either the information received by the structure MetaObject is used as the source of the actions to be ordered.
- The information is used by the structure MetaObject to check consistency and completeness of the actions with those of the action MetaObject.

This dual interpretation is generally useful in that it permits the completion of the MetaObject specifications to be performed sequentially or concurrently Objects. Other models include the Entity's actions and the Entity's structure models (Fig. 5). The mapping can be interpreted as a recipient of two or one ways (Fig. 4):

Fig. 4 MetaObject template for system specification diagrams

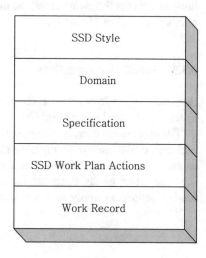

SSD Style

Domain

Specification

SSD Work Plan Actions

Work Record

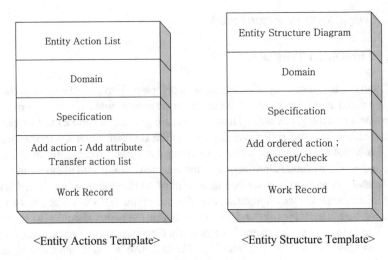

<Entity Actions Template>   <Entity Structure Template>

Fig. 5 MetaObject template for entity and entity structure

- Or an information received by the used metaobject structure is as a source of ordered actions.
- Come information is used by the metaobject structure for a verification of consistence and integrity of actions with the metaobject action [12].

3.3.2 MetaObject Hierarchies

A compact formula can be defined for each entity that combines these plans and indicates the transmission of information between them. In this case, there is a single unit of each one of the sub-templates. In general, only one templates indicates sub-template and interactions, but the number of instances may well depend on the particular demand and consequently, on instantiation. This situation is illustrated in the following level, the number of entity plan instances depends on the number of entities identified.

3.3.3 Specific Constructions

A system specification is a configuration of specifications given in selected instances of MetaObject (Fig. 6).

The application domain scheme provides a variable dimension to the means. For each part of the domain (entity), the model can necessarily create a configuration of MetaObjects. Figure 7, for the initial steps of the model, with a possible diagram of the description of the configuration as the formal part of the workplan. The following definitions apply to a configuration language with the allowed changes.

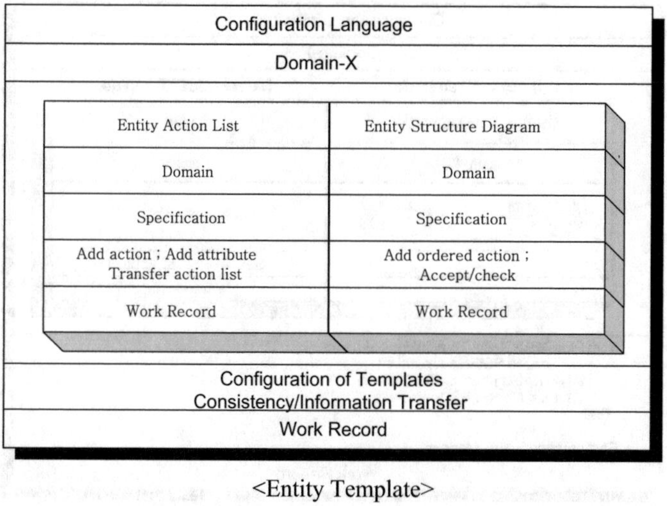

<Entity Template>

Fig. 6 MetaObject template for system specification diagrams

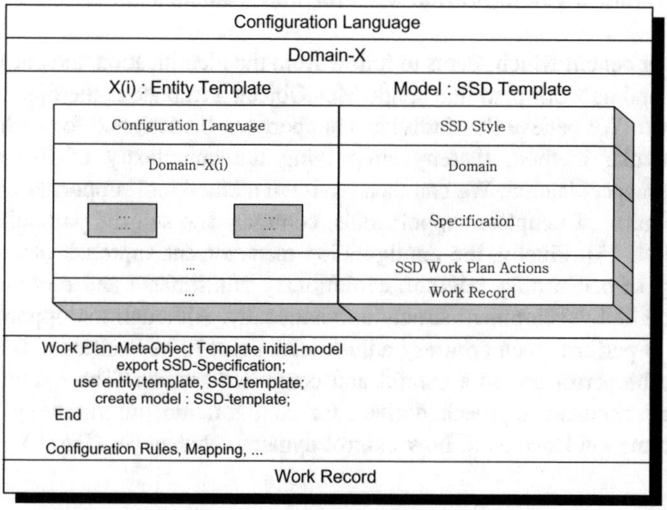

Fig. 7 Initial model specification using MetaObjects

The model is the possibility of the tool holder. We believe that the individual carrier can be designed on each model of a particular method, which simplified complexity metaobjets specific. We can consider is provided a set of plate of method configures model tool with the particular successive method's adopted [13].

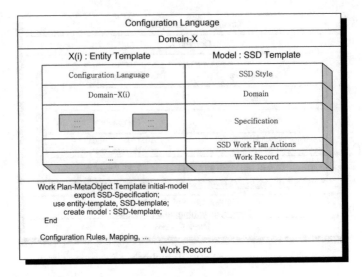

Fig. 8 Initial model specification using MetaObjects

3.3.4 Graphical Configuration Monitor and Management

A particular benefit which seems to follow from the identification and encapsulation of a style and its work plan in a single MetaObject. Template is the opportunity for tool support. We believe that individual support can be designed for each template in a particular method, thereby simplifying the complexity of that particular MetaObject specification. We can then envisage method tool support as comprising a configuration of template support tools, configured to suit the particular method adopted [14, 15]. Finally, the configuration management approach also offers the possibility of performing external, evolutionary adjustments and modifications to the method and development structure dynamically. Although the opportunity and facilities to perform such arbitrary adjustments seem to be desirable. They should obviously be performed in a careful and controlled manner. The specific kind of change management approach derived for configuration programming may also provide some guidance as to how control dynamic change [9] (Fig. 8).

4 Conclusions

The program configuration, with the use of a separate configuration language is a great way to express the structure of the system rather than the integration of structural choices in the same software components. This approach system gener-ates understandable, manageable and subject to change and facilitated the com-mission of software tools to support the construction and management of the system. The environment for distributed programming is an example of configuration for programming. In this paper, we discussed a configurable part, similar to that of the programming configuration, can be combined with the classification of meta-objects to provide similar benefits to the problem of integrating the instrument method. It was a proposed vision of configurable framework.

The meta-objects approach to software development supports the use of multiple meta-objects to partition the domain information, the method of development and formal representations used to express the software specifications, specific system and processes the configuration of the related meta-objects. The partitioning of knowledge shown in the approach facilitates this meta-object distributed development and the use of multiple systems of is presented.

Acknowledgements This work is supported by Catholic University of Daegu, Republic of Korea. This Research was supported by the International Research & Development Program of the National Research Foundation of Korea (NRF) funded by the Ministry of Science, ICT & Future Planning (Grant Number: K 2014075112).

This research was supported by the MSIP (Ministry of Science, ICT and Future Planning), Korea, under the ITRC (Information Technology Research Center) support program (IITP-2017-2013-0-00877) supervised by the IITP (Institute for Information & communications Technology Promotion).

References

1. Zhang, D., Jangam, A., Zhou, L., Yakut, I.: Context-aware multimedia content adaptation for mobile web. Int. J. Networked Distrib. Comput. (2014)
2. Yamada, A., Mizuno, O.: Classification of bug injected and fixed changes using a text discriminator. Int. J. Softw. Innov. (IJSI) **3**(1), 50–62 (2015)
3. Councill, B., Heineman, G.T.: Definition of a software component and its elements. In: Component-Based Software Engineering: Putting the Pieces Together, pp. 5–19 (2001)
4. Kim, HK.: Integration for Mobile Development Tools and Methods
5. Kang, M., Choi, O., Shin, J., Baik, J.: Improvement of software reliability estimation accuracy with consideration of failure removal effort. Int. J. Netw. Distrib. Comput. **1**(1), 25–36 (2013)
6. Chen, S., Sun, D., Miao, H.: The influence of alias and references escape on Java program analysis. In: Software Engineering Research, Management and Applications, pp. 99–111. Springer International Publishing (2015)
7. Park, J., Seo, Y.S., Baik, J.: A comparative analysis of reliability assessment methods for web-based software. Int. J. Softw. Innovat. (IJSI) **1**(3), 34–47 (2013)
8. Bendix, L., et al.: Software configuration management in software and hypermedia engineering: a survey. Handb. Softw. Eng. Knowl. Eng. **1**, 523–548 (2001)
9. Kramer, J., Finkelstein, A.: A configurable framework for method and tool integration. Softw. Develop. Environ. CASE Technol. 233–257 (1991)
10. Disterer, G., Kleiner, C.: Using mobile devices with BYOD. Int. J. Web Portals. **5**(4), 33–45 (2013)
11. Smith, B.C.: Procedural reflection in programming languages (Doctoral dissertation, Massachusetts Institute of Technology). Retrieved 16 Dec 2013
12. Laender, A.H.F., Liddle, S.W., Storey, V.: Conceptual Modeling-ER 2000: 19th International Conference on Conceptual Modeling, Salt Lake City, Utah, USA, 9–12 Oct, 2000 Proceedings., Springer, 31 Jul 2003

13. Nagappan, M., Shihab, E.: Future trends in software engineering research for mobile apps. In: Software analysis, evolution, and reengineering (SANER), 2016 IEEE 23rd International Conference on, vol. 5, pp. 21–32. IEEE (2016)
14. Park, J., Seo, Y.S., Baik, J.: A comparative analysis of reliability assessment methods for web-based software. Int. J. Softw. Innov. (IJSI). **1**(3), 34–47 (2013)
15. Zhang, J., Hu, G., Lee, R.: Formal specification and implementation of priority queue with starvation handling. In: Computer and Information Science 2011, pp. 155–167. Springer, Berlin Heidelberg (2011)

Author Index

© Springer International Publishing AG 2018   203
R. Lee (ed.), *Applied Computing & Information Technology*,
Studies in Computational Intelligence 727, DOI 10.1007/978-3-319-64051-8

Printed in the United States
By Bookmasters